CHINA 2020

OLD AGE SECURITY

CHINA 2020 SERIES

China 2020:
Development Challenges in the New Century

Clear Water, Blue Skies:
China's Environment in the New Century

At China's Table:
Food Security Options

Financing Health Care:
Issues and Options for China

Sharing Rising Incomes:
Disparities in China

Old Age Security:
Pension Reform in China

China Engaged:
Integration with the Global Economy

 THE WORLD BANK
WASHINGTON D.C.

OLD AGE SECURITY

PENSION REFORM

IN CHINA

 THE WORLD BANK
WASHINGTON D.C.

Contents

This report uses Hong Kong when referring to the Hong Kong
Special Administrative Region, People's Republic of China.

Acknowledgments

his book is based on the findings of a World Bank mission that visited China during August 1995. The mission members were Ramgopal Agarwala (team leader and task manager), Estelle James, Yan Wang, Sonia Xiaosong Zhao, Monika Queisser, and Cheikh Kane (World Bank staff), and Barry Friedman and Wu Xiaoyong (consultants). Rajiv Nundy (World Bank staff) developed the models for the quantitative analysis in the report, and Zhou Xiaoji (consultant) assisted. The counterpart agency from the government of China was the State Planning Commission, and the counterpart team consisted of Yao Hong and Wang Jinduo. The mission received valuable assistance from officials of the Ministries of Finance, Labor, and Civil Affairs; the State Commission for Restructuring the Economic System; the People's Bank of China; the People's Insurance Company of China; the All-China Federation of Trade Unions; and several line ministries in Beijing as well as in the provinces and municipalities

(Beijing; Shenyang, Liaoning Province; Changchun, Jilin Province; Kunming, Yunnan Province; Guangzhou, Guangdong Province; Fuzhou, Fujian Province; Nanjing, Jiangsu Province; and Zhengzhou, Henan Province).

Under the leadership of the counterpart team, thirty-four background papers were prepared by local authorities (provincial and municipal) and sectoral ministries presenting the current situation on their pension systems, as well as current plans and longer-term prospects. These background papers provided the raw material for the study. The report also drew on background papers prepared by Chinese and foreign experts, including Dong Wenjie, Guo Shuqing, Kang Huaiyu, Wang Yanhua, Yao Hong, Yu Mingde, Zuo Xuejin, Mukul Asher, Santiago Plant, and Wang Shaoguang.

Many experts inside and outside the World Bank also made contributions. Among these were Nicholas Barr, Emmanuel Jimenez, Klaus Schmidt-Hebbel, and Dimitri Vittas (peer reviewers); Chingboon Lee (UNDP Beijing); J.V. Gruat (ILO Beijing); Songsu Choi, Alan Gelb, Athar Husain, Vikram Nehru, Bertrand Renaud, and Jed Shilling. Rhoda Blade-Charest, Leila Cruz, Meredith Dearborn, Joan Grigsby, Guo Huiying, Jean Ponchamni, Jennifer Solotaroff, and Cherry Wu helped with production. The report was prepared under the general guidance of Pieter Bottelier, Nicholas Hope, Richard Newfarmer, and Michael Walton.

The book was edited by Meta de Coquereaumont, designed by Kim Bieler, and laid out by Glenn McGrath of the American Writing Division of Communications Development Incorporated.

China's current pension system has two severe problems: the urgent and immediate problem of the pension burden of state-owned enterprises, and the longer-term problem arising from the rapid aging of the population. The current system is incapable of tackling either problem. It also fails to contribute to economic development.

China today has a golden opportunity to move toward a sustainable, unified pension system that combines a defined benefit basic public pillar with funded individual accounts. This report argues that China's transition costs are lower, and its capacity to bear these costs higher, than in other countries that have made similar transitions because of four factors: the benefits to be reaped from unification of the pension system, rapid economic growth, structural change in the economy, and expected capital gains. But these four factors will remain operative only for the next fifteen to twenty years.

This book recommends a unified pension system that includes both mandatory funded individual accounts and a social insurance scheme. In addition, the financial implications for individual pensioners and for the pension system as a whole are simulated, showing how the proposal's financial viability will depend on demographic and macroeconomic developments and pension coverage. The book endorses a sustainable contribution rate that attaches major importance to long-term financial viability (more than sixty years). Risks associated with low compliance rates and low interest rates, among others, are also examined. Pension reform will fail if the system's coverage cannot be extended, compliance rates decline, or financial sector reform and capital market development do not provide an adequate return on pension reserves.

A careful program of consensus building along with mobilization of funds for implementation is needed for successful pension reform. Reform will also require accompanying reforms in legal, administrative, and financial systems. Moreover, decentralized public and private pension institutions should be allowed to compete on an equal footing, and investment rules for pension funds should encourage diversification and forge a stronger regulatory framework.

Overview

China's pension system has two severe problems: the long-term problem of a rapidly aging population and the immediate and urgent problem of funding pensions for employees of state-owned enterprises.

First, the long-term problem. China's population is aging rapidly, a process accelerated by the one-child policy of the late 1970s and the 1980s and increased life expectancy. By 2030 the absolute size of the labor force in China will begin to decline, and by 2050 the ratio of workers to pensioners (age sixty-five and above) is projected to decline to about 3 to 1 from the 10 to 1 ratio in 1995 (figure 1). The number of the elderly will rise from about 76 million in 1995 to 300 million by 2050.

The immediate and urgent problem is the pension crisis in the state-owned enterprise sector. State enterprises inherited heavy pension obligations from the central planning era. With the transition to a market economy, employment in the state enterprise sector is declining, while the number of pensioners is rising rapidly. In some cases the ratio of pensioners to workers is over 100 percent. An associated problem is the slowing of restructuring. The bankruptcy or sale of a state enterprise raises the difficult issue of how the commitment to pensioners (along with other social welfare obligations of enterprise) will be honored. When alternative arrangements for pensions and other social services are not available, enterprise reform is halted in its tracks since liquidations, joint ventures, or mergers cannot proceed smoothly until the social obligations of state-owned enterprises are assigned elsewhere.

Key features and shortcomings of the current system

The formal pension system in China is a largely urban-based, pay as you go, defined benefit system that covers mainly the state sector in urban areas. The nonstate sector, which now accounts for more than half the employment in many localities, has only spotty coverage. Though localities are trying to bring the nonstate sector under the formal pension system, most nonstate firms are resisting because current contribution rates are so high and the benefits so uncertain.

For thirteen provinces and twelve municipalities for which data are available, the simple average of contribution rates in state enterprises in 1994 was 23.5 percent for the provinces and 25.9 percent for municipalities, well above the international norm. The range is quite wide: from 19 percent in Guangdong to 28 percent in Henan. For eleven sectors that are exempted from municipal pension pools and are allowed to have their own pension pools, the average is just 15.9 percent, with lows of 10–15 percent for civil aviation, construction, banking, electric power, and petroleum and natural gas and a high of 24.5 percent for coal mining.

Noncompliance and exemptions account for much of the disparity between costs and contributions needed to break even. Moreover, as contribution rates have risen, compliance rates have been declining. Many municipalities have reported a drop in compliance rates from 90 percent in the early 1990s to 70 or 80 percent in the first half of 1995.

Responsibility for paying pension bills has been shifted from individual enterprises to groups of enterprises at the county, municipality, or prefecture level. This pooling of responsibility is intended to spread risk and help ensure payments to pensioners from enterprises with heavy financial burdens. Many localities have separate pools for different ownership forms. Provincial pooling has also begun in nine of the country's thirty-two provinces, but enterprises generally remain responsible for record-keeping and for delivery of pension benefits. And in many places pooling is only partial: enterprises with larger proportions of retirees in the pooled system have a higher contribution rate than those with lower proportions of retirees. Thus the pooling system has made only partial inroads into the individual-enterprise based system.

Economic and financial shortcomings of the current system

The current pension system can deal with neither the short-term problem of pensions in state-owned enterprises nor the long-term problem of old age security in China. The current system is financially unsustainable and fails to contribute to reform of state-owned enterprises or to the economic development of the country.
- It fails to solve pension problems of state enterprises because pooling is limited, noncompliance and exemptions are high, and the dynamic nonstate sector is not covered.
- It fails to solve the long-term problem of old-age security because the system's partial coverage means

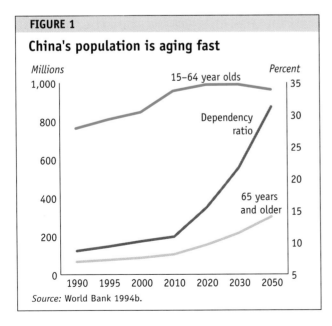

FIGURE 1

China's population is aging fast

Source: World Bank 1994b.

that a majority of China's old people will have no pension when they retire. Only small pension reserves have been accumulated in most municipalities, and these reserves suffer from poor management and earn a low rate of return.

- It fails to contribute to economic development in key ways, as specified below.

Fails to delink social welfare from enterprise management. Enterprises still carry a large share of responsibilities for their retirees. In many cases an enterprise's contribution rate depends in part on its own pension obligations. And in most cases enterprises keep pension records, pay pension benefits, and take care of pensioners' social service needs, including housing and health care. Thus enterprises carry a heavy burden of administration for the elderly, and alternative arrangements will have to be made if nonviable enterprises are to be liquidated. Partial reform created a new municipal-level bureaucracy without achieving any significant economies of scale in administration or relieving enterprises of responsibility for the elderly.

Fails to establish a level playing field. Similar enterprises in two provinces may have to pay widely different payroll taxes that can range from 20 to 30 percent. An enterprise may lose competitiveness not because efficiency in its core business is low but because it is situated in a locality with many retirees.

Impedes labor mobility. Factors of production must be able to move from one sector or region to another if China is to establish a socialist market economy and restructure its state enterprises. If a nonviable mill closes down in one city, laid-off workers need to be able to move to areas where similar industries might be expanding. For that to happen, the pension benefits of workers must be portable. The division of the national system into many separate unfunded municipal pools makes portability difficult and will become an increasingly serious impediment to labor mobility. Because individual accounts are largely notional, departing workers cannot take their personal accounts with them.

Fails to increase capital accumulation. The individual accounts being set up under the current system contain few if any assets because almost all incoming revenues are being used to pay current obligations to pensioners. Interest rates paid on account balances are also notional. Total pension reserves were estimated at less than 1 percent of GDP in 1995, and they are not increasing in many localities. Such notional accounts fail to meet the prefunding, or capital accumulation, objectives of pension funds. When workers retire and start drawing annuities based on their individual accounts, the annuities will have to be financed on a pay as you go basis out of contemporaneous contributions, which will have to rise dramatically to meet pension obligations.

Leads to inefficient capital allocation. Present regulations require that 80 percent of pension funds be invested in government bonds and the rest be kept in bank balances. Since the rates of interest set by the government have been below the inflation rate in recent years, these reserves lost value over time. Thus, even these small reserves will not be there to help when the population ages. What little discretion is allowed in investment portfolios is exercised by local officials who control the reserves and invest them locally, even if this does not maximize returns. Thus the opportunity to allocate pension capital to the most productive uses is lost.

Misses the opportunity for term transformation. Funded pension funds can create an enormous pool of resources for long-term investments in domestic infrastructure and other projects. The current system of unfunded notional accounts, combined with the absence of reliable long-term financial instruments, loses out on the opportunity for term transformation of savings. China has enormous needs for infrastructure and other long-term investments. Demand for infrastructure investment alone is projected to be as high as $744 billion for 1995–2004, or 7.4 percent of GDP (World Bank 1995d). But because of the absence of long-term instruments, most household savings are in short- and medium-term deposits, which do not provide a solid basis for long-term lending. Thus, foreign financial resources are often mobilized for infrastructure investments (often at high guaranteed rates of return), while domestic savings are inefficiently utilized.

The system is unsustainable

The Chinese official team analyzing the pension burden over the long term (1995–2050) came to the following dramatic conclusions:
- By 2033, if the pay as you go system is not reformed, it will require a contribution rate of 39.27 percent.
- A fully funded system with an additional contribution rate to meet the needs of transition cohorts will require a 34 percent contribution rate from 2004 to 2031.
- A combination of social pool and individual accounts can smooth the rise in contribution rates, but it will still require a 28 percent contribution rate for 2001–2050 if borrowing (from future generations or other sources) is not used.

The situation will be even worse for localities with high dependency ratios. Officially projected contribution rates for 2020 are 38 percent for Tianjin and 42 percent for Shenyang. Clearly, such rates will make most enterprises nonviable. The regional disparities in contribution rates that these scenarios imply will also be economically and socially unacceptable. Some innovative solution to the problem must be found.

Principles for a reformed system

The problems of the current system of pensions in China are deep-seated and widely recognized. Policy analysts have engaged in extensive discussions about how to solve these problems and design a new, sustainable pension system. The broad directions of reform are becoming clear, but implementation has proven difficult, hampered by misperceptions about the costs of transition and by the political implications of the change.

Unifying the system

The government has recently decided to unify the pension system. The program calls for the four unifications—a "unified system, unified standards, unified management, and unified fund usage"—which are to apply to all types of enterprises and workers. Enterprises and workers covered under separate plans or not covered at all would be brought into a single system with common standards. Management of the program would be transferred from enterprises to government agencies, and administrative management and fund management would be separated. There would be multiple channels of funding, including contributions from workers and employers. The idea of multiple tiers of benefits was reaffirmed, including supplementary benefits from enterprises and individual savings.

Although the target is a unified system by the turn of the century, the coexistence of two possible models has created a host of new problems. The two plans both involve individual accounts and social pooling, although organized and combined in different ways, and no final decision has been reached on a unified basic benefit tier. Plan I, based on ideas developed by the State Commission for Restructuring Economic Systems, emphasizes individual accounts, while plan II, originating with the Ministry of Labor, emphasizes a larger social component. Proposing two models and allowing localities to choose between them or to mix and match have created a hornet's nest. Each municipality attempts to differentiate its scheme from others in order to retain control over the pension system and its funds. This has led to the creation of hundreds of schemes across the country.

One way to resolve this dilemma is to make a sharp distinction between the mandatory and voluntary parts of the pension system, and between the overall structure of the system and its detailed features. The basic structure of the mandatory part needs to be unified, while specific regional or occupational needs can be met through special features and the supplementary part. This is consistent with practices in most industrial countries that specify the structure of mandatory systems on a nationally uniform basis.

Reducing the pension burden

There is consensus about the need to reduce the pension burden, but no agreement about how to do so effectively. Three variables can be adjusted to reduce pension burdens: retirement age, wage replacement rate, and pension indexation. The retirement age (60 for men and 55 for women) should be raised in accord with longer life expectancy. Today, the average duration of retirement is over sixteen years. Both to ensure the viability of the pension system and to avoid losing skilled personnel to early retirement, the retirement age should be gradually increased to 65—and even higher as longevity continues to increase. Since retirement age is a key variable affecting system finance, the sooner China moves to a higher retirement age, the sooner can contribution rates be lowered for a given level of benefit.

Wage replacement rates in China's pension systems are high by international standards and compared with Chinese practice before the Cultural Revolution. Replacement rates for cash wages are over 80 percent, and in-kind benefits (housing, medical benefits, and so on) continue at the preretirement level, resulting in a replacement rate of 80 to 90 percent of total compensation. Replacement rates in most countries are 40 to 60 percent (they were 50 to 70 percent in China before the Cultural Revolution). There is a broad consensus that the replacement rate should gradually be brought down to about 60 percent of net wages (wages minus payroll contributions from workers).

Pensions should be indexed to prices rather than wages. Currently, most municipalities index pensions to nominal wages, with an indexation coefficient of 40 to 80 percent. This is a very uncertain form of indexation, leading in some cases to a reduction in the real value of pensions and in others to a real increase, depending on the rate of inflation. A more appropriate adjustment mechanism, in line with international practice and the stated objectives in many localities, is full indexation to the consumer price index (CPI).

Introducing funded individual accounts

A substantial part of retirement income should come from fully funded individual accounts. The advantages of individual accounts for a rapidly aging population are well known. The pay as you go system runs the risk of requiring very high contribution rates by 2030, when economic growth rates in China may be slowing down. This may lead to high rates of evasion by enterprises, a high burden on society, and large intergenerational transfers. Funded individual accounts might avoid these problems and instill a spirit of self-help among workers who would look after their own retirement costs rather than depending on contributions from their children (whether in the context of the family, enterprise, or social transfers). Individual accounts also develop

incentives for workers to save more and retire later and to see that enterprises are making their contributions and that pension fund management companies are maximizing rates of return.

Notional individual accounts do not meet all these individual and social objectives. The current notional accounts link individual contributions and individual benefits, but the accounts do not accumulate reserves, do not earn an adequate real rate of return, and do not achieve the term transformation of savings. Thus contribution rates will have to rise dramatically to pay for promised pensions as the population ages. Many Chinese policymakers recognize these shortcomings but have not yet devised a practical way to fund individual accounts and pay current pensioners at the same time.

Ensuring an adequate real rate of return on pension funds

Funded individual accounts will not yield acceptable wage replacement rates for pensioners unless the real rate of return on pension funds is at least equal to the real growth rate of wages. If the real interest rate is zero and real wage growth is 5 percent annually, a 10 percent annual contribution rate will yield a wage replacement rate of only 10 percent (see box 1.2 in chapter 1). Most calculations of replacement rates made by official agencies assume that the rate of return equals the wage growth rate, yet the actual rate of return has been far lower than the wage growth rate for many reasons, including irrational interest rates, fraud, and poor management.

The rate of return on pensions can be high only if the general interest rate structure is rationalized, a reform that will increase the financial costs of state enterprises and the government. A short-term solution is to offer government guarantees on minimum real rates of return on pension funds. A longer-term solution would be to rationalize interest rates, decentralize management of pension funds, and allow diversification of investments, with the objective of allocating capital to the highest productive uses. Given the high productivity of capital in infrastructure and other long-term investments in China, and the high expected returns already paid to foreign investors, it should be possible to devise a scheme that allows pension funds to earn substantially higher yields (chapter 4).

Limited reform is risky

Pension system simulation results using an actuarial model developed for this report confirm the finding of Chinese analysts that the current defined benefit system is not financially viable over the long term. Design changes such as extending coverage, indexing to prices (but not to wages), reducing replacement rates, and increasing retirement age can improve the financial viability of the system. However, the pension funds would remain exposed to high risks (such as lower compliance and labor force participation rates and lower rates of return) if the defined benefit system continues. In the downside scenario the required contribution rates will rise to about 30 percent, well above international standards. It is desirable to reduce these risks (by improving compliance rates and rates of return on pension funds) and share them between society and individuals. That points in the direction of a multipillar system that combines social pooling with individual accounts, the conclusion reached by Chinese policymakers.

The key issues concern the specifics of the pillars and the transition to the multipillar system. The report uses a simulation model to test the implications of variants of such a multipillar system and suggests a preferred approach.

Moving toward a three-pillar system

The preferred approach is a three-pillared pension system based on the directions of reform identified above and recapped here:

• A unified structure for the mandatory part of the pension system, with flexibility through supplementary pensions.
• A substantial part (more than half) of pensions to come from individual accounts, which will be fully funded.
• A publicly managed part of the system to provide social insurance.
• A gradual move toward a target replacement rate for pensions of about 60 percent of net wages during the specified preretirement period.
• A unified retirement age for men and women that is gradually raised to 65.
• Indexation of pensions to the consumer price index.
• Pension obligations to current pensioners and work-

ers under the old system to be honored and financed through a combination of sources that do not impose an undue burden on the present generation nor any burden on the budget beyond the contingent liability of a mandatory pension system.
• Reform of the financial sector so as to maximize the rate of return on pension funds and allocate capital to the highest-productivity uses.
• Gradual expansion of coverage of the pension system to all formal sector workers in urban areas (including civil service and public institutions) and employees in large township enterprises.

The three pillars

The proposed system combines social pooling with funded individual accounts. It provides for a basic pension component to keep retirees above the poverty line (pillar 1) and for large mandatory individual accounts (pillar 2) to obtain the advantages of funding, supplemented by voluntary accounts (pillar 3) as desired. A transition mechanism will settle the implicit pension debt to current pensioners and workers. Pillar 1 thus has a redistributive element, since keeping the elderly above the poverty line involves some redistribution from higher to lower wage earners. Pillars 2 and 3 handle savings and link benefits closely to contributions. Pillars 1 and 2 are mandatory, while pillar 3 is based on voluntary private insurance or annuities. The transition mechanism will vanish in the very long run but is a key issue in the short and medium run (figure 2 provides details about the objectives, management, payments, and funding of the three pillars).

Pillar 1. The first pillar, intended as a redistributive and social insurance scheme, would provide a basic benefit equal to 0.6 percent of the local (provincial, urban, or rural) average wage per year of covered service. A worker with forty years of covered service earning an average income would get a wage replacement rate of 24 percent from the basic pension. The replacement rate would be higher for a low-income worker and lower for a high-income worker. This basic pension would help to equalize the incomes of the elderly. It would initially be financed through enterprise contributions of 9 percent of the wage bill. This low rate is made possible by expanding coverage to sectors that are not

FIGURE 2

Objectives, management, and funding of the proposed three pillars

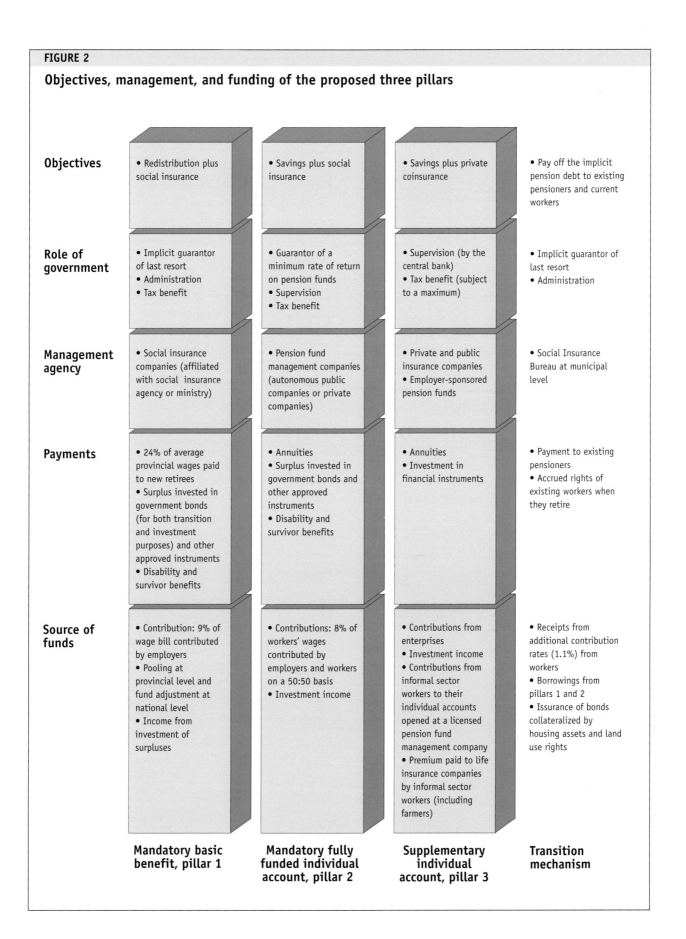

	Mandatory basic benefit, pillar 1	Mandatory fully funded individual account, pillar 2	Supplementary individual account, pillar 3	Transition mechanism
Objectives	• Redistribution plus social insurance	• Savings plus social insurance	• Savings plus private coinsurance	• Pay off the implicit pension debt to existing pensioners and current workers
Role of government	• Implicit guarantor of last resort • Administration • Tax benefit	• Guarantor of a minimum rate of return on pension funds • Supervision • Tax benefit	• Supervision (by the central bank) • Tax benefit (subject to a maximum)	• Implicit guarantor of last resort • Administration
Management agency	• Social insurance companies (affiliated with social insurance agency or ministry)	• Pension fund management companies (autonomous public companies or private companies)	• Private and public insurance companies • Employer-sponsored pension funds	• Social Insurance Bureau at municipal level
Payments	• 24% of average provincial wages paid to new retirees • Surplus invested in government bonds (for both transition and investment purposes) and other approved instruments • Disability and survivor benefits	• Annuities • Surplus invested in government bonds and other approved instruments • Disability and survivor benefits	• Annuities • Investment in financial instruments	• Payment to existing pensioners • Accrued rights of existing workers when they retire
Source of funds	• Contribution: 9% of wage bill contributed by employers • Pooling at provincial level and fund adjustment at national level • Income from investment of surpluses	• Contributions: 8% of workers' wages contributed by employers and workers on a 50:50 basis • Investment income	• Contributions from enterprises • Investment income • Contributions from informal sector workers to their individual accounts opened at a licensed pension fund management company • Premium paid to life insurance companies by informal sector workers (including farmers)	• Receipts from additional contribution rates (1.1%) from workers • Borrowings from pillars 1 and 2 • Issurance of bonds collateralized by housing assets and land use rights

now covered and that do not yet have retirees. In the initial years, a 9 percent contribution rate would generate some surplus, which could be used to finance part of the cost of transition from a pure pay as you go system to a partially funded system. There are, of course, many options for the design of pillar 1, and each is associated with risks and tradeoffs. These risks are examined in chapter 3.

Pillar 2. The second pillar would consist of mandatory individual accounts that would be fully funded and financed equally by workers and enterprises (with possible increases in workers' shares over time). The combined contribution rate would be 8 percent of a worker's individual wages. If the rate of return on pension funds equals the rate of wage growth, an 8 percent contribution rate would yield a replacement rate of about 35 percent of wages. The two mandatory pillars together would provide a target replacement rate of 60 percent for the average worker. A funded pension system, however designed, cannot achieve its objectives without a capital market and a financial system that pays a positive real rate of return to long-term savings. This report assumes that China will move swiftly to market-determined interest rates that remunerate savings.

Pillar 3. The third pillar would consist of supplementary pensions offered by employers on a voluntary basis or through individual accounts established by informal sector workers (including farmers) in licensed pension fund management companies or life insurance policies purchased from insurance companies. The amount would vary according to enterprise preferences and capacities and the willingness of informal sector workers to save for old-age security. Pillar 3 would be fully funded and portable.

Payments and funding

For an average worker who joins the new system, the target replacement rate after forty years of service would be about 60 percent of net wages during some specified preretirement period. An average of 24 percentage points of the 60 would come from pillar 1, and the rest from pillar 2. For a worker whose salary is 40 percent below average, the expected replacement rate would be about 80 percent, with 40 percentage points

coming from pillar 1. For a worker with wages 40 percent above average, the replacement rate would be about 51 percent, with 15 percentage points coming from pillar 1.

The mandatory component of contributions would be set at 17 percent of wages (9 percent for pillar 1 and 8 percent for pillar 2), which is below the current average rate of about 23 percent and in line with rates in such East Asian countries as Japan and Malaysia. Pillar 1, which would fund the basic pensions of new retirees, will have a surplus in current cash flows until 2031, with enough reserves to cover future deficits. Supplementary pensions would be financed on a voluntary basis by workers and employers. The transition would be financed by borrowing (through bonds) from pillars 1 and 2 in the short run and by a combination of modest additional contributions from workers and proceeds from the sale of state-owned enterprise assets in the long term (see below).

The government is the implicit guarantor of last resort for pillar 1 and the transition plan. As guarantor the government will play a key role in the administration of pillar 1 and the transition. The government's responsibility for pillar 2 would be as supervisor and the guarantor of certain financial instruments. For pillar 3, supervisory functions would be discharged by the central bank, as part of its role as supervisor of financial institutions.

Transition plan

Current pensioners will continue to receive their current benefits, fully indexed to the CPI. Current workers who retire after the new system is in place would receive the basic benefit, an annuity based on their individual accounts, an accrual rate equal to 1 percent (which could be slightly higher in the initial years)[1] of their salary for each year of service before reform, and supplementary pensions that their employers might have purchased.

Financing the transition

There is no consensus on how to finance the transition to a multipillar system. With funding, much of the current contributions of enterprises and workers would go to individual accounts. Other financial resources will be needed to pay the current pensioners and the accrued pension rights of workers who had contributed under the old system (the implicit pension debt). The problem is how to make these "double" payments.

Some rough calculations by Chinese policy analysts suggest that the implicit pension debt could be three to four times GDP. These high estimates have discouraged the authorities from explicitly recognizing these debts and reforming the system. Estimates of the implicit debt depend on the promised benefits of the pension system, the age structure and life expectancy of the working population and retirees, and the discount rate used. Reasonable values for these variables suggest that the implicit pension debt is actually less than 50 percent of current GDP and an even smaller proportion of expected GDP, given China's anticipated growth rate. This debt is smaller (largely because of the relatively low coverage rate of China's pension system today) than that for countries such as Chile and Argentina. This is an opportune time for China to move to a system with funded individual accounts. If China instead maintains its pay as you go system, its pension debt will grow rapidly as coverage expands.

Countries that have made the transition from a pay as you go to a funded scheme have used a mix of instruments to bridge the financing gap. The instruments available are the same as those for other public expenditures: bonds, higher taxes, lower spending, or transfers of public assets. The choice and mix of instruments will determine the impact of pension reform on economic growth and the distribution of the costs and benefits of transition between and within generations. Most reforming countries have borrowed to spread the burden across many generations and to smooth contribution rates over time. Some of the borrowing (through issuing marketable bonds) could come from the funds accumulating in the mandatory pillars. The bonds can be redeemed by future contributions or by other sources of revenue, such as proceeds from the sale of state-owned enterprise assets.

The report presents illustrative calculations to show the implications of various methods of financing the transition and proposes a burden-sharing formula. Under this formula current pensioners will be paid by an additional contribution rate levied on all covered workers. As noted in chapter 3 (table 3.9), this requires an additional contribution rate of 1.1 percent. For settling the pension debt of current workers, interest-earning and fully collateralized bonds worth 1,236 billion yuan would be given

to the transition agency. Up to 2010 the transition agency would borrow from pillars 1 and 2 to cover the payments to pensioners over and above the receipts from the additional 1.1 percent contribution rate. After 2010 the transition agency would start cashing in some bonds—say, 10 percent. These cash flows will be enough to make transition payments and start repaying the debt to pillars 1 and 2 (see table 3.9). By 2050 receipts from additional contribution rates will substantially exceed transition payments, and the loans to pillars 1 and 2 will be paid off soon thereafter. With this burden-sharing arrangement, the total mandatory contribution rate will become 18 percent, and transition bonds (for all covered workers including civil service and public institution workers) will be worth 1,236 billion yuan, which could be safely collateralized by the housing assets (and associated land-use rights) of the state sector.

China's window of opportunity

China's costs of transition to a funded system are lower than those in many other economies, and its financial capacity to fund the transition is greater, for at least four reasons: benefits from unification of the pension system, rapid growth of GDP, structural change in the economy, and expected capital gains. These four favorable factors will remain operative over the next fifteen to twenty years but will largely disappear after 2030. They cannot be counted on to finance high pension costs permanently, but they are well suited to financing the temporary needs of the pension transition:

• *Unification*. If the system is unified and coverage is extended to the nonstate sector (including township enterprises), substantial resources would be generated that could be used to fund the transition. Contribution rates would need to be kept low so as not to encourage evasion or discourage economic growth. Simulations show that it should be possible to keep rates low.

• *Growth*. China's economy is growing at more than 10 percent a year, and its domestic savings are over 40 percent of GDP. Such rapid growth makes it feasible to use a portion of the incremental income to fund the transition. An important mechanism for accomplishing this funding is to index benefits to prices rather than wages, to keep the rise in pension costs below the economic growth rate so that some of the benefits of growth can be siphoned off for pension reform.

• *Structural changes.* Associated with rapid economic growth are rapid structural changes in the economy: labor shifting from rural to urban areas, from agriculture to nonagricultural activities in rural areas, and from the state sector to the more efficient nonstate sector in urban areas. Even if the overall labor force grows slowly, the reserve pool of surplus labor will enable the nonagricultural labor force to expand rapidly, thereby expanding the contributory base for the pension system. Eventually, these workers too will retire. But if the transition has been made by then, their pension needs will be met in part out of their own funded individual accounts and in part by the higher productive capacity generated by their retirement savings.

• *Capital gains.* Inevitably, as state-owned enterprises are restructured and divested of such social responsibilities as housing, some of their assets will be sold. Some of the proceeds from asset sales could be used to redeem the bonds that are issued in the early years of the pension reform. The state enterprise sector in China, while facing problems, has not collapsed the way it has in other socialist economies in transition. The sector provides about 70 percent of government revenues and accounts for a significant part of the country's impressive savings, export, and growth performance. The value of state-owned enterprise assets is higher than GDP (compared with less than 50 percent in Latin American and Eastern European countries). The availability of these assets opens up possibilities for transition financing in China that are not available in these other economies. Two principles might be honored in this process: any asset transferred to pension funds has to be marketable and carry a value corresponding to the market-determined prices, and the link between asset sales and pension liabilities should be as close as possible, so that the asset-liability swap is easy to understand and implement.

Managing the reformed system

Successful implementation of the proposed pension system depends on accompanying reforms in legal, administrative, and financial systems. Reforms in all these areas must go hand in hand, and coordination is crucial. To ensure that pension reform moves forward across the country, a national agency should be created, with primary responsibility for coordinating the development of the three pillars and the transition mechanism (see chapter 4).

Administering the first pillar

The large diversity in initial conditions and administrative capacity may preclude national pooling in the short run; thus the initial pooling for pillar 1 should probably take place at the provincial level, with regional disparities evened out somewhat through a national adjustment fund. Pooling at the provincial rather than municipal level will provide a larger pool of workers and pensioners to spread the risk. It will also facilitate the incorporation of township enterprises into the system, which is beyond the authority of most municipal authorities. Over time the system should move toward national pooling, with a uniform national contribution rate for the basic benefit.

Particular attention should be paid to boosting compliance. Improved tax administration capacity should reduce evasion, but the likely growth of the informal sector will increase it. One of the best ways to increase compliance is to create appropriate incentives and tax enforcement mechanisms at the design stage.

Managing the second pillar

The fully funded individual accounts should be decentrally and competitively managed by licensed corporatized state, joint venture, or private investment companies, subject to rules governing minimum capital, reserves, investment allocation, information disclosure, and fraud. To prevent individual accounts from becoming notional, they must be managed separately from the first pillar and the transitional mechanism. Any borrowing by the transition agency should be explicit, using fully collateralized bonds with a positive real rate of interest and ensured repayment.

In the short run it is likely that most pension funds would be invested in public securities. The government should guarantee a minimum real rate of return on pension funds. Funds mobilized in this way should be used for long-term capital construction or infrastructure projects with an economic rate of return that justifies the real cost of capital.

Over the longer term strategies should be developed to encourage greater competition in management and to permit investment diversification across sectors (public and private), across financial instruments (equities, bonds, and mortgages), and across geographic regions. Decentralization and diversification are important for allowing the second pillar to maximize returns (subject to an acceptable risk level), to encourage the best allocation of capital, and to assist in the financial market development needed for China's economic development. To this end the government should formulate investment regulations and adjust them periodically as capital markets and real estate markets mature and grow. These regulations should preclude concentration in one company, industry, or locality.

Developing and regulating the voluntary third pillar

China needs to develop its institutional and regulatory capacity for providers of supplementary pensions: insurance companies and employer-sponsored pension funds. Two major impediments are the weak incentives on the demand side due to the high wage replacement rate under the old pension system, and the lack of a legal framework for employer-sponsored pension funds and for pension fund management companies. The recommended reduction in the replacement rate should create space for voluntary plans, and a legal framework should be established to regulate the plans. In addition, informal sector workers (including farmers) should be allowed to establish individual accounts in licensed pension fund management companies or purchase life insurance policies and enjoy tax benefits for pension contributions up to a stipulated maximum amount. Furthermore, competition should be introduced into the insurance industry by restructuring the People's Insurance Company of China and by encouraging new domestic companies and foreign and joint venture companies. Clear and transparent investment rules should be set up for insurance companies, similar to those for pension funds (see table 4.2). More opportunities should be allowed for portfolio diversification and capital appreciation.

Administering the transition mechanism

A special administrative arrangement would be set up in each municipality to pay off the pension obligations owed to current pensioners and workers under the old system, to finance the transition to a new system, and to meet cash flow needs by borrowing from pillars 1 and 2 through bond issuances and redemptions.

Conclusions

Pension system reform involves a major redirection of financial flows and asset entitlements and affects the livelihoods of millions of people at a highly vulnerable stage of their life. In the process, there would be major gainers and losers. Enterprises and localities with younger populations would lose initially, and those with older populations would gain through the pooling proposed here. The proposed scheme reduces conflict by emphasizing individual accounts and defined contributions. However, the basic pension component would have a redistributive element. And the transition plan may involve major changes in the entitlement to assets in society. The treatment of women would also change. Women would get access to a basic benefit that is independent of their wage, a gain in most cases since women tend to be low wage-earners. However, they lose from the delay in their retirement age to match that of men. Pension reforms thus impinge on many powerful political considerations, which may play a decisive role in policymaking but are beyond the scope of this report.

Pension system reform in China is urgent. The urgency comes primarily from the need to delink social welfare responsibilities from state enterprise management so as to accelerate reform of the sector. From the long-run point of view it is clear that the costs of transition will rise with time. But that does not mean that reform should take place in an atmosphere of crisis. The many risks involved in reform must also be considered. If the financial sector reforms and capital market reforms do not materialize, the individual account system will not function properly. Particularly important are interest rates that provide positive real returns to savings. Similarly, if coverage of the system cannot be extended as expected, compliance rates decline, or local authorities do not cooperate fully, the reforms will fail.

A careful program of consensus building along with mobilization of funds for implementation is needed for successful reforms. Passage of a Social Insurance Law is urgently needed to standardize the basic framework of pension provision and to establish the institutional infrastructure and regulatory framework of the system. Experiments with the standardized system in selected localities should be started soon, along with careful monitoring of the results. These results should be used to improve and implement the unified system of pensions by 2000 at the latest.

Note

1. A gradualist scenario was considered with an accrual rate of 1.2 percent, to be reduced to 1 percent over twenty years; postponement of the increase in retirement age to 2010; and 10 percent (instead of 20 percent) coverage of township and village enterprises, rising to 50 percent by 2020 (instead of 2010). The model indicates that the long-term financial viability of the system is not affected. However, the transition debt increases by 120 billion yuan, and the surpluses in pillars 1 and 2 after funding transition payments are lower by about 16 billion yuan a year in the first ten years. The ability of social insurance system savings to fund infrastructure and other investments is thus reduced. Political leaders need to decide the balance between the higher economic costs of a more gradual transition and its greater political and social acceptability.

chapter one

The Dual Challenges of the Pension System

China's program of economic reforms since 1978 has achieved spectacular successes. While most socialist economies in transition in Eastern Europe have been traumatized by shock therapies, China has registered GDP growth of more than 10 percent a year over the past fifteen years, without a single year of decline in output. Its inflation rate has averaged less than 10 percent a year, though it was as high as 25 percent in one year. With a savings rate of more than 40 percent of GDP, export growth exceeding 15 percent a year in current dollars, and foreign exchange reserves of over $100 billion, China's economic position is strong. This economic situation helps position China to bear the costs of transition, including pension reform.

China's economic success has its roots in many factors—economic, social, and historical. A notable part of that success is the capacity of policymakers to focus intensively on problems as the occasion demands, mobilize domestic policy analysis, solicit foreign advice, form political consensus, and then move decisively on policy reforms. Inflation control and reforms in the fiscal sector, financial sector, and foreign exchange regime show this process at work.

Pension system reform is one of the areas on which the Chinese authorities are focusing attention. Enormous amounts of analytical work have been done by Chinese policy analysts, numerous experiments are being conducted across the country, and various sources of foreign advice are being sought. This volume is a contribution to that discussion, which is now at a stage where decisive action is possible—and strongly recommended.

The challenges

China has two challenges of old age security: a long-term problem rooted in the rapid aging of its population, and the immediate and urgent problem of pensions for employees of state enterprises.

The long-run problem

The one-child policy of the late 1970s and the 1980s and increasing life expectancy mean that China's population is aging rapidly. By 2020 the elderly (age sixty and older) will make up some 16 percent of the population, close to the 18.6 percent share in OECD counties in 1990 (table 1.1). It took most OECD countries 80 to 100 years to double the proportion of their old people to 18 percent; China will do that in just 34 years. By 2020 China's 16 percent share of elderly will be much higher than that of its neighbors: for example, 6.3 percent in Pakistan, 7.6 percent in Bangladesh, 9.0 per-

cent in Vietnam, 10.3 percent in India, 10.9 percent in Indonesia, and 12.8 percent in Thailand. The government projects an even more rapid rise in urban areas, with estimated dependency rates climbing from 14.8 percent in 1990 to 47.3 percent in 2030. The crisis is dramatized by the oft-repeated observation in China that when the people currently entering the work force retire, one couple will have to support four parents: the "1–2–4" phenomenon—one child, two parents, four grandparents.

Thus in 2030 China will be facing the problems of a mature economy like Japan while its per capita income will, in all probability, be about one-fifth that of industrial countries today. Moreover, just as China's old-age burden increases, its GDP growth rates may be slowing as opportunities for technological catch-up diminish. In other words, China will have a high-income economy's old-age burden with a middle-income economy's resources for shouldering it. Early planning is needed to avert the crisis by using the opportunity provided by today's high growth and high savings rates to prepare for old-age security tomorrow.

Short-term challenges

A more urgent and immediate problem is the pension crisis in the state enterprise sector. As a legacy of the plan era, state-owned enterprises have heavy pension obligations (table 1.2). With marketization of the economy, employment growth in the state enterprise sector is slowing and the number of pensioners relative to employees is rising—in some cases to more than 100 percent. Under central planning the profits of enterprises were pooled and resources were allocated according to plan, so the pension burden in individual

---

TABLE 1.1

Current and projected share of the elderly in the population
(percent)

Country	1990	2000	2010	2020	2030	2050
China	8.9	10.2	12.0	16.0	21.9	26.1
India	6.9	7.5	8.3	10.3	13.1	20.4
Korea, Rep. of	7.7	10.7	13.9	19.5	25.5	29.8
Malaysia	5.7	6.5	8.0	11.0	14.5	22.1
Japan	17.3	22.7	29.0	31.4	33.0	34.4
OECD average (simple)	18.6	20.0	23.2	26.9	30.8	31.3

Note: The elderly are defined as people sixty years and older.
Source: World Bank 1994a.

enterprises was not their concern. Today, individual enterprises are becoming responsible for their profits and losses, and many are unable to manage the burden of pensions. Even when an enterprise's core business is profitable, it may show losses because it has a large pool of pensioners. Its capacity to operate, borrow, and expand may suffer. Many pensioners may not get adequate payments simply because of their past assignment to enterprises that are now doing poorly financially. Thus the system is becoming both inefficient and unfair.

An associated problem is that the pension system is slowing enterprise restructuring. State-owned enterprises are like miniwelfare states, with work units that look after workers' welfare "from cradle to grave."[1] The bankruptcy or sale of a state enterprise raises the difficult issue of how the commitment to pensioners will be honored (along with the other social welfare obligations of the enterprise). When alternative arrangements are unavailable, the result is a slowing of enterprise reform. Liquidation, joint ventures, and mergers cannot proceed smoothly unless the issue of the social obligations of state enterprises is settled. In addition, as the losses of enterprises have accumulated, banks now face an increasing burden of bad debts, which impedes their own commercialization.

Thus the dilemma of policymakers is increasingly evident: state enterprises cannot be efficient unless they face a hard budget constraint; banks cannot be commercialized unless they can enforce hard budget constraints on their borrowers; and neither of these is possible unless enterprises are freed of their social welfare obligations, of which pensions are an important component. As the losses of state enterprises mount, so does the urgency of solving their pension problems.

The authorities are well aware of the seriousness of both the short- and the long-term problems of the pension system. After ten years of experimentation throughout China and an enormous amount of analytical and policy work, the time has come to take stock of the results and to take decisive action on reform, much as was done in reform of taxation and foreign exchange.

Evolution of the old-age pension system

A brief history of the modern pension system in China puts the current discussion in perspective. The first formal pension system was established in 1951 through the State Council's Regulations on Labor Insurance. The regulations applied to enterprises with more than 100 workers. Not surprisingly, it was a small program because the enterprise sector was small and there were few retirees. One year after the start of the program, there were only 8 million enterprise workers and 20,000 retirees, or more than 400 workers per retiree. The system was funded exclusively by contributions from enterprises. A contribution rate of 3 percent of the wage bill was sufficient to finance the system on a largely pay as you go basis. Old age pensions covered 50 to 70 percent of workers' wages. Seventy percent of enterprise contributions were retained locally to pay pensions while 30 percent were transferred to a national master fund or pools, an arrangement that seems to have provided a degree of prefunding. The All-China Federation of Trade Unions managed both the local payment procedures and the master fund.

Subsequent reforms expanded coverage. The Temporary Regulations on the Retirement of Employees in Government of 1955 set up a separate system for employees of government units, nonprofit units, and party organizations. The Temporary Regulations on the Retirement of Workers and Staff of 1958 dealt with both

TABLE 1.2
Growing imbalance between workers and pensioners in state-owned and collective enterprises

Indicator	1990	1991	1992	1993	1994	1995
Workers (millions)	106.8	109.6	108.7	109.0	105.5	103.2
Pensioners (millions)	13.2	14.7	15.9	17.7	19.1	22.4
System dependency ratio (percent)	12.4	13.4	14.6	16.2	18.1	21.7
Wage expenditure (billions of yuan)	223.5	251.3	291.5	351.9	448.8	533.7
Pension expenditure (billions of yuan)	22.6	26.6	32.5	42.0	56.5	70.7
Pension-wage ratio (percent)	10.1	10.6	11.1	11.9	12.6	13.2

Source: China, Ministry of Labor.

enterprise and government workers. Coverage was extended to enterprises with fewer than 100 workers. This system continued until the Cultural Revolution, which began in 1966.

During the Cultural Revolution the All-China Federation of Trade Unions and the Ministry of Labor were abolished. The pension funds that had accumulated were used for other purposes, thus eliminating any prefunding that had built up. Supervisory responsibilities were transferred to local labor bureaus, while responsibility for managing payments was transferred to enterprises. Pooling ended and so did prefunding, since each enterprise paid the pensions of its own workers out of current revenues.

Reforms of the system in the 1970s

When the economic reforms began in 1978, the State Council issued new pension regulations (Document 104) for state-owned enterprises, government workers, and nonprofit organizations and recommended that large collectively owned enterprises follow the same rules. The prevailing retirement ages were reaffirmed— 60 for men and 55 for women—with additional adjustments for hazardous jobs. Qualifications were eased, allowing a worker to retire after ten years of continuous service rather than twenty. New, higher benefits were related to length of service and to the final standard wage. Someone who had worked for at least twenty continuous years would get a pension of 75 percent of the standard wage, someone who had worked for fifteen to twenty years would get a 70 percent pension, and someone who had worked for ten to fifteen years would get a 60 percent pension. There was a minimum guaranteed pension of 30 yuan per month. Disability pensions were related to the final standard wage and the extent of care needed.

The regulations of 1978 created a number of problems that subsequently had to be addressed. The regulations were clearly intended to encourage early retirement in order to create jobs for a large influx of new workers into the urban labor force. Among the incentives were increased retirement benefits, a lowering of the minimum years of service required to qualify for retirement, and for a time, the guarantee of a job for one child following a worker's retirement. The number of retirees jumped fivefold between 1978 and 1985, and pension costs rose from 2.8 percent of the urban wage bill to 10.6 percent. As the increase in pension costs became apparent, the incentives to encourage retirement were curtailed.

Other problems have persisted longer. Basing the pension on the final standard wage has provided an incentive for workers and employers to jack up the final wage in preparation for retirement. The 1978 regulations provided a generous replacement rate relative to the standard wage, which was the largest part of the wage at that time. Labor reforms have since increased bonuses to the extent that the standard wage is perhaps only half the total wage compensation. While this should have reduced the effective replacement rate and thus helped control the rise of pension costs, actual benefits may be more generous than the rules suggest.

Reforms in the 1980s

New regulations have been issued several times since 1986, and there have been many experiments in pension design. State Council Document 77 of 1986 established pooling across state enterprises on a limited basis at the municipal level. Enterprises retained responsibility for distributing pensions. The pool operated by setting a contribution rate (or formula) for participating enterprises. If the pension costs of an enterprise were less than the contribution rate, the difference was remitted to the pool. If pension costs were higher, the pool would cover the difference.

The 1986 pension reforms were accompanied by employment reforms that established contract labor. New workers were to be hired on a contract basis, while current workers would continue as permanent workers. Separate city pension pools were established for contract workers and permanent workers. Contract workers made individual contributions, while permanent workers initially did not. Enterprises contributed to both pools. In the late 1980s pooling was extended to workers in collectively owned enterprises in many cities. Since then, other enterprise types, such as joint ventures, joint stock companies, and foreign enterprises, have been brought into pension pools in some cities, although participation is generally far from complete. Some provinces are moving toward provincial-level pooling, generally only for state enterprise workers, but participation remains fragmentary.

State Council Document 33 of 1991 called for individual contributions by all workers, in addition to enterprise contributions, and for experiments including a role for individual accounts. It also called for an expansion of pooling and the establishment of three tiers in the pension system: a basic benefit, a supplementary benefit to be provided by enterprises in sound financial condition, and a benefit based on individual saving.

Goals of the current reform

The government has articulated the goal of establishing a unified pension system by 2000. Enterprises and workers now covered under separate plans or not at all would be brought into a single system, with multiple channels of funding, including contributions from workers and employers. Management responsibilities would be transferred from enterprises to government agencies, although post offices and banks might assist with payment procedures. Administration and fund management would be separate. The idea of multiple tiers of benefits was reaffirmed, including the enterprise supplementary benefits and individual savings tier.

Although the target is a unified system by the turn of the century, for now State Council Document 6 of 1995 proposes two models for the basic tier. City and prefecture governments were given the right to select a reform design and provincial governments the right to approve or disapprove the choice. While a final decision has not yet been reached on a unified basic benefit tier, both plans involve individual accounts and social pooling, although organized and combined in different ways. Plan I, based on ideas developed by the State Commission for Restructuring of Economic Systems, emphasizes individual accounts, while plan II, based on the ideas of the Ministry of Labor, emphasizes the social component.

The government may have inadvertently contributed to further fragmentation of the system by proposing two plans. By allowing localities to chose between them or any combination of components of the two, it has opened up a Pandora's Box. Each municipality is attempting to differentiate its scheme from that of others in order to retain control over the pension system and the pension funds. This has led to creation of hundreds of schemes all over the country.

Plan I

In plan I the basic pension system for new workers would be individual accounts. A social pool would be responsible for pensions for those who are already retired, for current workers not fully covered by individual accounts, and for certain adjustments for retirees drawing from individual accounts. Contributions into individual accounts, at approximately 16 percent of total wages, would consist of three parts:
• An individual contribution of 3 percent of total wages.
• An enterprise contribution of 8 percent of each worker's total wage.
• An enterprise contribution of 5 percent of the average local wage.

The intention is to increase the individual contribution over time and to decrease the enterprise contribution (by 1 percentage point every two years for ten years) until individuals are contributing half the total to their individual accounts. Both a ceiling and a floor are set for the individual contribution. For the ceiling, the wage base used in calculating contributions will not include individual wages in excess of 200 percent or 300 percent (to be set by each city) of the average local wage. For the floor, every worker will have to contribute at least 60 percent of the average city wage, no matter how low the actual wage. This minimum contribution will in turn establish a floor benefit upon retirement. The local government may set a higher floor for retirees and supplement the pension so that it reaches the floor. Interest will be credited to individual accounts each year, based on the bank interest rate and the rate of increase in the average local wage, though the exact mechanics for determining the rate of interest are not clear.

A worker who reaches retirement age after contributing for at least fifteen years will receive a monthly pension equal to 1/120 of the total accumulated in the individual account. With a life expectancy of sixteen years at age sixty, this implies an interest rate of 4.5 percent. The social pool will continue to pay pensions for people who live longer than expected. A worker with less than fifteen years of contributions will receive the amount in the individual account in one lump sum on retirement. If a person dies, the accumulation based on individual contributions will be paid to heirs, and the part paid by the enterprise will be returned to the social pool. The total enterprise contribution rate will be larger than the approximately 13 percent of wages going into individual accounts, in order to fund the social pool. Each city will set its own contribution rate. Any shortfalls in the social pool will be made up through withdrawals from the individual accounts. To the extent that funds are withdrawn to meet shorfalls in the pool, the individual accounts will be notional rather than fully funded.

Plan I includes two government guarantees: a longevity guarantee and a minimum pension guarantee. Workers who outlive their individual accounts will continue to receive a pension through the social insurance pool. Workers who contribute to the new system for at least fifteen years or who have had a continuous employment tenure (including the contribution years) of at least ten years before the reform will receive a monthly pension after retirement. If the pension is less than the minimum pension (specified by local governments), the government (the level is not specified) will make up the difference.

Plan II

Plan II puts more emphasis on social pooling than on individual accounts. The plan is designed mainly for cities that had chosen an earlier Ministry of Labor pension model, with a vesting period of ten years. For those whose contribution period is longer than ten years, the pension will consist of the following parts:
- A social pension equivalent to 20–25 percent of the local average wage.
- A premium pension equivalent to 1.0–1.4 percent of the wage base for each year of contribution.
- An individual account pension that can be drawn as a lump-sum or an annuity equivalent to the funds in the individual account.
- A supplementary subsidy from the social pool, which will be eliminated over time.

As in plan I transitional arrangements will cover those who have already retired and the cohorts of workers not fully covered by the new scheme. The floor and ceiling for contributions will be as in plan I, except that earnings in excess of 300 percent of the average local wage will be excluded from the wage base.

A modified plan IIB eliminates the premium pension over time, leaving just the social pension and individual accounts, which will be gradually increased. The increase in individual accounts and the reduction in the premium pension will be coordinated in order to maintain a 60 percent wage replacement rate for the total pension during the transition.

Key problems of the current system

The current pension system suffers from a variety of shortcomings, from spotty coverage and inadequate scope for pooling and portability to high levels of instability and uncertainty.

Low and variable coverage

China's formal pension system is largely urban-based. In rural areas extended families are the chief means of support for the elderly (see box 1.1). This arrangement is often associated with the "son preference" of the farmers, which often comes into conflict with the government's one-child policy, especially if the first-born is a girl. Many policymakers in China feel that better pension provision in rural areas can improve compliance with the one-child policy and alleviate its worst side effects. A small beginning in rural pensions has been made by the Ministry of Civil Affairs, though the voluntary scheme is too small to provide adequate protection for the elderly (see box 1.1). It is difficult to provide formal old age insurance for farmers, who have small, fluctuating incomes and poor recording systems. But as China industrializes and the labor force in agriculture declines over the long term, a formal pension system can be extended to rural areas.

Even in urban areas pension system coverage has focused largely on the state sector. Coverage of the nonstate sector, which in many localities now accounts for over 50 percent of employment, has been spotty, ranging from 20 to 90 percent. While most localities are trying to bring the nonstate sector under the pension system, the high contribution rates and uncertain benefit system are provoking resistance from most nonstate sector firms. The nonstate sector is the most dynamic part of the economy, so the local authorities should proceed cautiously to avoid making nonstate firms uncompetitive through high and arbitrary taxation. Only with a rational and realistic pension system in place can the nonstate sector be persuaded to become full participants in the formal pension system.

High and variable contribution rates

Pension expenditures in China are only about 12.6 percent of the wage bill in the covered sectors, and surpluses are less than 0.5 percent of the wage bill. Thus effective contribution rates are about 13 percent of the wage bill. However, for many state-owned enterprises, the contribution rates are considerably higher.[2] Simple average contribution rates in 1994 were 23.5 percent for thirteen provinces and 25.9 percent for twelve

Rural pensions

While most urban workers in China have long been covered by pension plans offered by their enterprises, rural workers have had little access to formal pension plans and have instead relied primarily on the extended family for old age support. Parents had several children and invested in them, hoping for future returns. Multigenerational households were the rule, mobility was limited, and strong social norms reinforced reliance on the family as a social insurance and informal pension system. Because these forces are generally stronger in rural than in urban areas, formal systems of old age security made their first appearance in cities, in China as elsewhere.

In recent years, however, the government has become concerned about the adequacy of the informal system. Families are shrinking in size, and workers are becoming more mobile. In 1991 the Ministry of Civil Affairs introduced a voluntary pension insurance system for farmers and workers in town and village enterprises. The plan is now being tried on an experimental basis in 1,400 counties across the country. So far, about 50 million people are participating in this scheme and 4 billion yuan have been accumulated. Participation is highest in the richer rural areas along the coast and nearby.

The plan encourages workers to contribute voluntarily to their retirement savings accounts. The accounts are fully funded and are turned over to county officials to invest. Some counties invest in treasury bonds or bank deposits that earn 10 to 13 percent while others use a special trust arrangement with a bank that has greater flexibility and earns as much as 18 percent (nominal). Each year the individual accounts are credited with interest payments at a rate set by the ministry; currently, the nominal rate is 12 percent. In some years the rates earned by the fund and credited to individual accounts have been less than the rate of inflation.

Ordinarily, the money cannot be withdrawn until retirement at age sixty, when it is turned into an annuity that lasts a person's lifetime. An annuity can also be purchased with a lump-sum payment at retirement. Currently 510,000 people are receiving pensions, most of them on the basis of lump-sum payments. Annuities are based on a nominal interest rate of 8.8 percent and an expected lifetime of eighteen years after retirement; annuities are not indexed for inflation.

Why have so many peasants and workers chosen to join, despite the negative real interest rate and the loss of access to their money until retirement? One reason is that few alternative financial instruments are available. Most bank deposits earn only 10 percent, so the plan's 12 percent interest is a better deal. Another reason is that local officials who sponsor the plans and want them to work use education and moral suasion. In a suburb of Yantai, Shandong Province, with a participation rate of 87 percent, contributions are solicited during a fund-raising week held once each year, an administratively efficient arrangement that makes it easy for local officials to apply moral suasion. Local officials have constructed a table that shows workers exactly what size pension they will get from alternative amounts of contributions over their working lives. These pensions look quite attractive to most workers, especially since the numbers are not corrected for inflation. Township and village enterprises are supposed to use their tax resources or profits to partially match individual contributions. While only the more affluent villages and towns (such as Yantai) actually engage in matching, where it occurs it clearly raises the rate of return and consequently the incentive to participate.

Does this mean that voluntary pension plans are the way to go and that these plans should be provided by a government agency? A closer look at the results of the current scheme shows some flaws. While many individuals are participating, they are contributing minuscule amounts. In 1995, after three years of operation, the average accumulation per participant across all 1,400 counties was 80 yuan. At only about 2 percent of a farmer's annual income, this amount is unlikely to generate more than a 5 percent replacement rate of income after retirement.

How can this be corrected? First, private insurance and investment companies should be allowed to operate in rural areas, subject to regulation, and to offer annuities and long-term savings opportunities but without any implicit or explicit state guarantee. Second, if the extended family is breaking down even in rural areas, and if a sizable group of rural workers seems unprepared to make alternative arrangements on their own, mandatory coverage should be applied, but only after the mandatory system has been reformed and is no longer building up a pay as you go pension debt.

Workers in township and village enterprises that exceed a specified size, such as 50 or 100 workers, might soon be included in a reformed mandatory system that combines individual accounts with a social safety net. These enterprises are the fastest growing in the economy, employing about half the nonagricultural labor force in 1993. In the very long run, as the government's tax-collecting capacity grows, and if the extended family continues to break down, coverage can be extended to farmers, workers in smaller enterprises, and the self-employed groups that present the greatest administrative costs and monitoring problems in all countries.

municipalities for which data are available, well above the international norm of about 20 percent (table 1.3).[3] The range is quite wide as well: from 19 percent in Guangdong and 21 percent in Shenzhen to 28 percent in Henan and 30 percent in Chongqing. Contribution rates are quite a bit lower for eleven sectors that have their own pension pools. The average is 15.9 percent, with lows of 10 to 15 percent for civil aviation, con-struction, banking, electric power, and petroleum and natural gas, and a high of 24.5 percent for coal mining. These sectors account for 15 million of the 75 million state enterprise workers.

Moreover, effective contribution rates can also vary because of differences in the base used to calculate con-tributions. For example, in Shenyang enterprise contri-butions are 18 percent of wages plus 40 percent of the

TABLE 1.3

Features of pension schemes of selected provinces, municipalities, and sectors, 1994
(percent)

Locality	Covered pensioners/ Covered contributors	Average pension contribution rate			Wage replacement rate	Gross pension/ gross wage
		Employer	Worker	Combined		
Provinces						
Liaoning	27.2	21.9	3.0	24.9	—	21.9
Jilin	23.2	23.0	2.0	25.0	—	15.4
Heilongjiang	22.6	23.0	2.0	25.0	90	—
Jiangsu	21.7	20.0	3.0	23.0	65	14.0
Zhejiang	20.4	22.5	3.0	25.5	80	—
Fujian	22.7	20.0	4.0	24.0	81	13.5
Henan	16.3	25.0	3.0	28.0	75	17.5
Hubei	18.4	20.0	2.0	22.0	85	17.9
Guangdong	23.2	17.0	2.0	19.0	70	—
Hainan	26.9	18.0	3.0	21.0	65	9.9
Sichuan	25.4	19.5	3.0	22.5	95	22.3
Yunnan	26.9	21.0	3.0	24.0	71	22.3
Shanxi	—	20.0	2.0	22.0	95	—
Simple average	*22.9*	*20.8*	*2.7*	*23.5*	*79*	*17.2*
Municipalities						
Beijing	36.0	22.0	5.0	27.0	75	18.0
Tianjin	32.5	25.0	4.0	29.0	—	—
Shanghai	40.5	21.0	4.0	25.0	—	—
Harbin	26.2	22.0	2.0	24.0	102	21.3
Changchun	27.7	22.0	2.0	24.0	75	—
Shenyang	38.5	22.0	3.0	25.0	83	—
Taiyuan	24.5	20.0	3.0	23.0	100	—
Wuhan	36.9	26.0	3.0	29.0	90	—
Chengdu	14.0	22.0	5.5	27.5	83	—
Chongqing	33.1	27.0	3.0	30.0	60	—
Guangzhou	—	24.0	2.0	26.0	—	—
Shenzhen	3.3	13.5	7.5	21.0	61	—
Simple average	*28.5*	*22.2*	*3.7*	*25.9*	*81*	*—*
Sectors						
Civil aviation	5.9	8.0	2.0	10.0	—	—
Coal mining	28.7	21.5	3.0	24.5	88	25.0
Communication	—	—	—	—	—	—
Construction	19.7	10.0	3.0	13.0	83	—
Banking	10.7	12.0	2.0	14.0		—
Electrical Power	—	13.0	2.0	15.0	80	—
Hydropower	—	18.0	1.0	19.0	—	—
Nonferrous metal	29.3	—	—	—	—	26.8
Petroleum & natural gas	13.9	12.0	3.0	15.0	85	—
Railway	29.4	—	—	17.0	—	—
Transportation	—	—	2.0	—	80	—
Simple average	*19.7*	*13.5*	*2.3*	*15.9*	*83*	*—*

— Not available.
Source: Background materials provided by the provinces, municipalities, and relevant ministries.

pension expenditures of the enterprise, for an average effective contribution rate of 31.2 percent. In Jilin province the enterprise contribution rate is 21.5 percent of both wages and pensions, pushing the effective contribution rate as high as 50 percent for enterprises with high pension bills.

With contribution rates already high and rising compliance rates have been falling. Many municipalities and provinces reported compliance rates of 70 or 80 percent in 1994 and the first half of 1995, down from 90 percent in the early 1990s (table 1.4). The financial difficulties of an increasing number of state enterprises are to blame in many cases. In Jiangsu, a prosperous coastal province, 40 percent of state enterprises are running at a loss, and the situation is worse in other regions. Many of these state enterprises have negotiated delays in pension contributions, but it is not clear how many will resume paying. Another reason for worsening compliance is the lack of a legal framework and enforcement power of local social insurance agencies. China has no social insurance law, and contributions are consequently defined by provincial and local regulations, rather than by law. The social insurance agencies at the provincial and local levels lack the legal power to enforce payments or to take noncontributing enterprises to court.

TABLE 1.4
Reported compliance rates in selected municipalities
(percent)

Municipality or province	Contribution rates[a] (enterprises, workers)	Compliance rates[b]	
		Earlier years	Current
Beijing	Various (19–27, 5)	95	95
Tianjin	Various (20–30, 4)	—	95
Shanghai	30 (average 25.5, 4)	—	90
Shengyang	21 (18, 3)	—	80
Changchun	23.5 (21.5, 2)	91	76.9
Nanjing	21.5 (18.5, 3)	80–90	80
Wuhan	29 (26, 3)	—	90
Taiyuan	Various (18~25, 3)	—	80
Chengdu	24 (22, 2)	—	80
Chongqing	30 (27, 3)	—	70.2
Fuzhou	Various (21–29, 4)	95	90
Guangzhou Declining	Various (21.5–24.5, 2–3)	96	
Hainan Province	21 (18, 3)	—	70

— Not available.
a. Contribution rates often vary, depending on ownership.
b. Percentage of enterprise contributing, as reported by local officials; may be overestimated since state enterprises in financial difficulty that have negotiated delayed payments might not be considered in noncompliance.
Source: Background papers and interviews with local officials during the mission.

Inadequate scope for pooling and portability

Currently, most pension pooling is conducted at the county, municipality, or prefecture level, and many localities have separate pools for state enterprises, collective enterprises, or other population groups (table 1.5). Nine of the thirty-two provinces have experimented with some form of pooling. Four of them—Hainan Province and the three provincial-level municipalities, Beijing, Tianjin, and Shanghai—have the most complete form of provincial pooling, with a uniform plan and contribution rate for all urban workers. Five other provinces have partial pooling, through a provincial readjustment fund or separation of state enterprises from other enterprises, but contribution rates differ across localities and ownership types. Enterprises generally remain responsible for record-keeping and delivery of pension benefits. In many places pooling is only partial: enterprises still pay a higher contribution if they have a larger proportion of retirees.

Portability of pension benefits is still uncommon and difficult. The cities of Shanghai, Changchun, and Chengdu allow portability among their work units, but many other cities do not. Several provinces also allow portability across work units within the pool (Liaoning, Jilin, Zhejiang, Fujian, Hubei, Guangdong, and Hainan), but others do not (Heilongjiang, Henan, and Sichuan), and there is practically no portability between provinces. Even where portability is allowed, it usually requires separate negotiations between the old and new employers or pools, and agreement is difficult in an environment where enterprises care about costs and profits.

High degree of fragmentation

Fragmentation in China's pension system follows from the wide dispersion of authority over pension policy and administration and reflects the variety and complexity of the country's economy. The Ministry of Labor is responsible for pension provision and administration for employees of public enterprises, the Ministry of Personnel oversees pension and other benefits for civil servants[4] and employees of social (non-profit) organizations, and the Ministry of Civil Affairs is in charge of social welfare programs, including supplementary pension schemes in rural areas. The System Reform Commission, State Planning Commission, and

TABLE 1.5
Levels of pension pooling in selected provinces and provincial-level municipalities

Province or municipality	Level of pooling	Separate pools by ownership	Pension administration
Beijing	Province-level municipality pooling	One pool but different contribution rates by ownership	Department of Labor
Tianjin	Province-level municipality pooling	One pool but different contribution rates	Department of Labor; Social Insurance Company
Shanxi	Province	For state enterprises only	Department of Labor in provincial government
Liaoning	City or county	Separate pools for state enterprises and collective enterprises	Unified administration for six cities; fragmented for eight
Jilin	Province, for state enterprises and joint ventures	Separate county pools for collective enterprises	Social Insurance Company
Heilongjiang	City or county	Separate pools for state enterprises and collective enterprises	Department of Labor
Shanghai	Province-level municipality pooling	Uniform contribution rate for all enterprises	Bureau of Social Insurance
Jiangsu	City or county, with province readjustment fund	One pool in each city or county	Bureau of Social Insurance and Department of Labor
Zhejiang	City or county	One pool for state enterprises and collective enterprises	Social Security Bureau at the local level
Fujian	Complete provincial pooling with unified fund in 1984–93; now a contractual provincial pooling with readjustment fund	One pool but different contribution rates for state enterprise and foreign-invested enterprises; separate pools at city level for collectively owned enterprises	In Luoyuan, a general social insurance company was in charge of all subpools
Henan	Provincial pooling for state enterprises and collective enterprises	Contribution rates differ by locality and ownership type	Bureau of Social Insurance and Department of Labor
Hubei	City or county	Pools cover most state enterprises and some collective enterprises	Department of Labor
Guangdong	City or prefecture level, with provincial readjustment fund	Unified pool at each city or prefecture; foreign-invested and private enterprises uncovered	Bureau of Social Insurance
Hainan	Province	Uniform contribution for all urban workers	Bureau of Social Security under Department of Labor
Sichuan	Province, with a pension pooling fund	Uniform at city and prefecture level, but not for the whole province	Department of Labor
Yunnan	City or county level	Separate pools for state enterprises and collective enterprises	Bureau of Social Insurance

Note: In many cases record-keeping and pension payments are administered by enterprises.
Source: Compiled based on data from provinces.

Ministry of Finance have also contributed to the design and implementation of various pension reform plans.

Another impediment to unification is the excessive dispersion of authority to lower levels of government. By allowing lower levels of government to select from plans I and II, the central government is relinquishing its authority in pension provision. Provincial and local governments have modified or combined elements of plans I and II, essentially making individual decisions on all system parameters of pension provision. Many localities intentionally differentiate their schemes from others and introduce nontransparency so as to retain authority over their own program and the surplus funds that are emerging in the short run. As a result, there are now hundreds of separate pension schemes in China.

Instability of the system

The frequent changes in basic policy and continuous experimentation with various pooling schemes have created uncertainties about future contribution and benefit rates that are adversely affect the economy at the individual, enterprise, and national level.

Pensioners and workers are uncertain about their livelihoods at a vulnerable stage of their life. Workers do not know what replacement rate they will get when they retire. Pensioners do not know whether their pensions will be fully indexed to wages or inflation.

The uncertainties about the pension system are also causing major planning problems for enterprises, particularly in the nonstate sector which cannot count on the state to bail them out of financial difficulty. Potential foreign investors will be discouraged not only by the high contribution rates but also by the uncertainty about what the rate will be in the future.

Pension system uncertainty is equally unsettling for municipalities and provinces, which will have to bear the burden of deficits if pension pools are not large enough to cover pension expenditures. And if reform goes the way of fully funded individual accounts (instead of notional accounts), localities may have large pools of pension funds to manage, which will require sophisticated fund management skills. Even at the national level the choice of pension system could have major implications for fiscal and monetary policies. Thus uncertainty about China's pension system is an impediment to forward-looking programs at all levels of society.

Adverse implications of the current system

The current pension system resolves neither the short-term problem of pensions in state enterprises nor the long-term problem of old age security in China. Nor is the system helping to accelerate state enterprise reforms or the economic development of the country.

The state enterprise problem remains unresolved

In the current system enterprises still carry a large part of the responsibility for their retirees. Enterprises keep records, pay pensions, and take care of pensioners' needs, including housing and health. The incomplete reforms in the pension system are creating a new bureaucracy without really achieving the objective of relieving enterprises of responsibility for the elderly.

Because of the fragmented system, state enterprises in some sectors (such as banking, civil aviation, and electric power) pay less than 15 percent of the wage bill in pension contributions, while many others pay more than 30 percent. The nonstate sector is inadequately covered and generally pays much lower contribution rates. Falling compliance rates are increasing the burden on state enterprises that do comply. Thus the fragmentation of the pension system and its uneven coverage are responsible for the high pension burdens of state enterprises. The present reform programs do not correct this basic problem.

The long-run pension problem remains serious

Pension funding is largely on a pay as you go basis, with little accumulation of reserves. The individual accounts that are being set up are largely notional (see box 2.2); they contain few if any assets. Thus when workers retire and start to draw on annuities based on their individual accounts, the annuities will have to be paid on a pay as you go basis, out of contemporaneous contributions. The contribution rate will have to rise steadily to meet those payments. These notional individual accounts will not solve the pension problem of the rapidly aging population.

With small beginnings in funded personal accounts and the necessary surpluses in operating accounts, total pension funds in the country were estimated to have about 44 billion yuan in reserves by the end of 1995. By regulation 80 percent of these funds must be invested in government bonds and the rest kept as bank balances. Rates of return on these assets have been below the inflation rate in recent years, which means that the reserves have lost value. Localities are ignoring the regulations and investing their funds in other projects. Apparently, only 50 percent of the reserves are invested in government bonds and bank accounts; the rest are invested in local projects, sometimes with much higher rates of return. Thus there is a strong incentive for localities to move toward nontransparent forms of investment and to keep resources in the localities, which may fail to maximize rates of return on a national basis. A more liberal system would encourage more transparent investment and maximize rates of return on a national basis.

The implications of the low rate of return on pension funds are serious for retirees who depend on their individual accounts. Differences in growth rates of real wages and interest can have an enormous impact. With a contribution rate of 10 percent of wages, the replacement rate will be over 40 percent if real interest rates and growth rates of real wages are both 8 percent, but the replacement rate will be only 10 percent if real wages grow at 5 percent and real interest rates are 0 percent (see box 1.2).

The government of China's Research Group for the Social Security System (China RGSSS 1995) came to the following dramatic conclusions about the long-term pension burden:
• A pay as you go system will require a contribution rate of 39.27 percent in 2033, when the dependency ratio reaches its peak.
• A fully funded system, with additional contribution rates for meeting the needs of retirees during the transition, will require a contribution rate of 34 percent from 2004 to 2031.
• The combination of a social pool and individual accounts can smooth the rise in contribution rates, but the rate will still remain high at 28 percent for 2001–2050.

BOX 1.2

Why it is important to have a high rate of return on pension funds

When a pension plan is funded, the accumulated funds earn a monetary return that helps to finance future pensions. In a market economy this financial advantage over pay as you go plans corresponds to the real return from the productive investments enabled by the pension funds, which increase economic growth. But problems arise if the rate of return is less than the rate of wage growth or inflation.

Suppose that wages in the economy are growing at 4 percent a year and that the wages of a typical worker are growing at another 1 percent because of increased experience and skills as the worker ages, for a total combined rate of 5 percent annually. Suppose further that the interest rate is 2 percent. In that case accumulated pension assets are not increasing as fast as wages and will not yield a high wage replacement rate when the worker retires. Under these circumstances a 10 percent contribution rate will yield only a 16 percent replacement rate (see box table). If a higher replacement rate, say 32 percent, is desired, the worker will have to save 20 percent of wages each year or the interest rate will have to rise to 5 percent.

In general, a pension system will be in financial trouble if the rate of return on its investments falls below the rate of wage growth (see the low replacement rates beneath the diagonally highlighted rates in the box table). Workers will be disappointed by low replacement rates, while the low return on investments indicates that the investments are not contributing much to economic growth.

The situation is even worse if the nominal interest rate is below the inflation rate. If inflation is 20 percent and the interest rate is 10 percent, next year the pension fund has only 92 percent as much purchasing power as it did this year. By the time the worker retires, pensions will be far less than wages and will purchase far less than they could have had the money been spent in earlier years.

If the interest rate is low because the fund is forced to lend to the government at specified rates, this constitutes a nontransparent tax on workers to finance government expenditures. Whether these expenditures were good or bad, whether they contributed to economic growth or not, this kind of pension plan financing makes no sense. If the interest rate is less than the rate of inflation a pay as you go scheme makes more sense.

Simple wage replacement rate with a 10 percent contribution rate

Real wage growth	Real interest rates			
	0	2	5	8
0	*22*	40	103	225
2	16	*27*	65	160
5	10	16	*34*	78
8	7	11	21	*43*

Note: This calculation assumes that an individual works for forty years and has twenty years of expected retirement. A zero inflation rate or a pension that is indexed for inflation is also assumed. Administrative costs are ignored. Real wage growth is economywide wage growth plus age-related earnings growth for the individual worker.

These projections fail to take into account the effect of rural-urban migration, but the basic conclusion about the nonviability of the current system is confirmed by more elaborate analysis (see chapter 3).

The situation is going to be worse for localities with high dependency ratios. The officially projected contribution rate in 2020 is 38 percent for Tianjin and 42 percent for Shenyang. Clearly, such rates will render most enterprises nonviable, particularly nonstate sector firms competing internationally and working under hard budget constraints. The regional disparities in contribution rates implied by these scenarios will also be economically and socially unacceptable. The present trends therefore cannot be allowed to continue. Some innovative solution to the problem must be found.

Loss of economic efficiency

In addition to failing to resolve either the short- or the long-term pension problems of the country, the current system results in a loss of economic efficiency.

Impediment to labor mobility. If China is to establish a socialist market economy, factors of production (including labor) must be able to move from one sector and region to another in response to price signals. This mobility is particularly important for making the restructuring of state enterprises feasible. If a steel mill or textile mill has to be closed down in one city, the redundant labor ought to be able to move to other municipalities, where similar industries might be expanding. Workers need to be able to carry their accumulated pension benefits with them. But since individual accounts are largely notional, there would be no real funds for departing workers to take with them, even if portability were permitted.

This is a serious handicap for state enterprise restructuring and industrial relocation, both priority issues in China today. The government's program to help the development of western provinces calls for some textile mills to be relocated from the coastal provinces to western provinces. While many of the workers for these mills will be found in the host provinces, much of the supervisory staff may have to come from the coastal provinces. If moving means that these workers lose their pension benefits, they will be unwilling to do so. Similarly, if the surplus workers are to move from the

northeastern to the southern provinces, the southern provinces would be reluctant to provide them with full pension benefits out of their resources. Only if pension rights are portable will workers, particularly those close to retirement, be able to move.

Absence of a level playing field. Under the current system two enterprises (for example, two electronics factories) in two provinces may have to pay widely different payroll taxes that could range from 20 to 30 percent. This is equivalent to imposing a value added tax at widely different rates for the same product. An enterprise might lose competitiveness not because its core efficiency is low but because it is in a locality with many retirees. Thus the system will end up allocating resources to enterprises that are not necessarily more efficient but that have the advantage of being located in an area with a younger population. China's northeast region, which has a high percentage of elderly, is already experiencing low growth. If it has to bear a higher tax for pensions, it will suffer a competitive disadvantage relative to other regions and its economic performance will lag further.

Missed opportunity for term transformation. One important objective of pension system reform is to help in the capital accumulation and productivity growth that drive an economy forward. China's current pension system does neither. Ultimately, an aging population will have to be supported by the output of the active working population. The pension system is important in determining the cost of these transfers from workers to retirees (whether through families, the government, or the pensioner's own account). Even more important is the effect of the pension system on the growth rate of the economy. Growth rates, compounded over the typical working life of forty years, can make an enormous difference to the output per worker. This influence works through the rate of saving in the economy and the productivity of capital.

China's savings rate is currently very high. However, with stabilization of GDP growth, the spread of conspicuous consumption, and the aging of the population, savings rates are likely to fall. A pension system with funded individual accounts can motivate households to save more, but this effect is muted in China's system with its small, notional individual accounts.

Even more serious are the missed opportunities for term transformation of savings. China has enormous needs for infrastructure and other long-term investments. Demand for infrastructure investment alone is projected to be as high as $744 billion for 1995–2004, or 7.4 percent of GDP (World Bank 1995d). Because of the dearth of long-term savings instruments, most household savings are in short- and medium-term deposits, which do not provide a solid basis for long-term lending. Thus foreign financial resources are often mobilized for infrastructure investments (often at high guaranteed rates of return), while domestic deposits are left underutilized. Funded pension funds can create an enormous pool of resources (4 to 6 percent of GDP) for long-term investments, thus supporting domestic infrastructure investment and a high rate of return for future pensioners. The current system of unfunded notional accounts is missing out on this opportunity.

Notes

1. Some amount of corporate welfare provision is not uncommon even in OECD countries. Companies often provide housing, medical facilities, and other forms of assistance to employees. The problem in China arises from the extent of these social welfare benefits, lack of alternatives for meeting these needs, and inadequate enterprise funding of these obligations.

2. One reason for these high rates is that they relate to previous years' wages.

3. Payroll taxes for pensions are below 20 percent in most countries in the world. In 1995 they were 12.4 percent in the United States, 18.6 percent in Germany, and 16.5 percent in Japan. Averages for developing countries in Latin America, the Middle East, Sub-Saharan Africa, and Asia, are about 10 to 12 percent. Singapore had an exceptionally high rate of 40 percent, but the funds were used mainly for housing loans by workers.

4. A common practice around the world.

chapter two

Key Issues and Options for Reform

The problems of the current pension system are widely recognized, and the broad directions of reform are becoming clear among policy analysts in China. Implementation has proven more difficult, hampered by misperceptions about the costs of transition and by the fact that change always involves losses for some groups. This chapter describes these directions of reform and argues that the costs of transition are manageable if China starts now.

The government's plan for pension system reforms calls for the "four unifications": equal treatment of all types of enterprises and workers by a unified system, unified standards, unified management, and unified fund usage. However, by proposing two models and allowing the localities to chose between them or any combination of their features, the program has led to a proliferation of schemes. The most challenging task of China's pension system reform is to reduce the fragmentation and make the system and its administration more unified.

Unifying the system

Most industrial countries specify the structure of the mandatory system on a nationally uniform basis, and most have unified publicly managed pay as you go pools, though there are some exceptions (see table A2.1 for description of the systems in various countries). But many started out with fragmented systems before moving to a unified one. Chile, Japan, the Republic of Korea, Mexico, the United Kingdom, and the United States all had fragmented pension systems in earlier periods. Japan had three public pension schemes prior to its 1985 reform: one for the private sector, one for the public sector, and one for farmers and the self-employed. Benefits varied considerably. The 1985 reform unified pension insurance and adopted a basic pension system with more equitable benefits and contributions. A transitional mechanism allowed employees to receive pensions from old systems while the new one was phased in.

It could be argued that the experience of other countries—smaller and more homogeneous—is not relevant for China, a vast country with large regional differences. But regional disparity is not unique to China; the United States and India also have significant regional differences (see table A2.2). Moreover, China has a unitary government rather than a federal one, and the legal powers of the central authority are more extensive than they are in a federal system (such as the United States). The Communist Party structure provides a powerful instrument for unification, when necessary. The introduction of the National Tax Service (breaking a centuries-long tradition of local-based tax collections) and of a center-local tax assignment system shows that the central authorities can move on unification when necessary. The need for unification of the pension system is all the stronger in China today, when there are signs of centrifugal tendencies in economic (and other) spheres.

With the pension system being extended to the non-state sector and contribution rates so high, compliance from enterprises (particularly private companies and joint ventures) will be difficult unless there is clear legal authority behind the system. The present effort at drafting a social security law has been frustrated by variable coverage and contribution rates across regions. Unification is essential for the establishment and implementation of a legal framework.

One way to resolve this dilemma is to make a sharp distinction between the mandatory and voluntary parts of the system and between its overall structure and its detailed features. The basic structure of the mandatory part of the system needs to be unified, while the details and the supplementary part can take into account specific regional or occupational needs. This is consistent with practices in most industrial countries, which specify the structure of mandatory systems on a nationally uniform basis. Chapter 3 gives some recommendations for China in this regard.

Reducing pension costs

Pension costs need to be reined in to make the system sustainable. Raising the retirement age, lowering the wage replacement rate, and indexing to prices rather than wages are the main steps for lowering costs.

Increasing the retirement age

Current retirement ages (60 for men and 55 for women) were set at a time when life expectancy was 50 years. Today it is 71 years. The retirement age ought to be gradually raised to 65 years, for men and women, both to reduce pension costs and to avoid the loss of skills to the economy through premature retirement. Recent surveys show that a significant percentage of retirees in fact continue to work, a finding in line with the global trend toward flexible schedules on retirement (see box 2.1).

The issue is controversial, however. Many state enterprises have surplus labor, and it is tempting to deal with it by retiring elderly workers early. But the cost to enterprises of retired workers is high. Better to extend the retirement age (thus saving on the pension burden) and to deal with labor redeployment and severance pay on a case by case basis. Often, it would be more cost-effective to reduce surplus labor by redeploying younger and middle-aged workers than by retiring older workers early.

Reducing the wage replacement rate

At more than 80 percent, wage replacement rates in China's pension systems are high by international standards. Add on the social benefits (housing, medical

Flexible retirement schedules

Labor force participation rates in common retirement age brackets show a decline from prime working age rates, but the rates vary broadly across industrialized countries. For men aged 55 to 59 participation rates range from 94.1 percent in Japan to 59.2 percent in the Netherlands. For women in that age group Sweden has the highest overall participation rate at 77.2 percent, but rates for women are as low as 20 percent in Austria, Ireland, Italy, the Netherlands, and Spain.

In the 60–64 age group the differences between countries become more clearly marked. The participation rate for men is over 75 percent in Japan but below 20 percent in Austria, France, and the Netherlands. For the over-65 age group participation rates are low for all countries, but the rate remains at over 50 percent in Japan and around 25 percent in Denmark, Ireland, Norway, Portugal, and the United States.

One of the first concerns of older workers is to ease the transition from work to retirement. Mandatory retirement or induced early retirement make for an abrupt transition. The exclusion of older workers in this way causes undue anxiety and can have negative effects on their health. A gradual reduction of working hours and transfer to lighter work are ways of easing the transition to retirement.

Phased retirement permits older workers to continue part time in their career job, quite different from seeking one of the part-time jobs open to older people in a precarious job market. Phased retirement has been successful in Sweden, but less successful elsewhere. One of the main reasons for a lack of success has been that the replacement income for the partial loss of earnings has been insufficient. In countries where pensions are based on the last year's earnings, workers are reluctant to take up phased retirement. In Sweden pensions are based on the best fifteen years of wages rather than on the last year. Elsewhere, as well, new attempts are being made to facilitate phased retirement. British Airways, for example, introduced phased retirement and overcame the "final salary" rules by allowing pension entitlements to be calculated on the basis of the full-time pro-rated salary.

Increasing the options for flexible retirement would go a long way toward erasing the social boundaries between workers and retirees and give older people the opportunity to remain economically active longer. In the long run it could help to reduce social tensions, as well as to keep pension system costs lower.

Labor force participation rates at retirement ages
(percent)

Country	Men, age groups			Women, age groups		
	55–59	60–64	65–69	55–59	60–64	65–69
Australia	71.8	48.7	8.4[a]	37.0	15.4	2.3[a]
Austria	63.0	12.7	3.6	23.8	5.2	1.2
Canada	73.6	47.6	16.6	47.5	24.8	7.8
Denmark	81.9	47.1	24.6	64.5	26.8	8.0
Finland	62.7	23.9	7.1	62.1	18.0	3.1
France	69.3	18.2	4.5	47.8	15.1	3.2
Germany	81.5	34.9	8.0	45.5	11.9	4.0
Ireland	79.8	59.4	26.5	22.1	14.0	5.9
Italy	68.9	37.2	12.6[b]	21.1	10.0	3.9[b]
Japan	94.1	75.6	55.3	56.4	40.1	28.0
Netherlands	59.2	18.0	—	19.3	4.2	—
New Zealand	80.3	39.0	8.7[a]	49.3	20.2	2.7[a]
Norway	81.8	61.5	25.8	61.5	45.8	16.7
Portugal	71.9	53.0	29.9	40.1	25.6	14.6
Spain	73.6	44.8	6.3	24.4	16.2	3.9
Sweden	82.5	57.8	—	77.2	49.1	—
United Kingdom	75.7	52.2	13.1	54.5	24.7	8.0
United States	77.4	54.9	25.8	57.0	37.4	16.0

— Not available.
a. Includes all people over the age of 65.
b. Includes people between the ages of 65 and 70.
Source: ILO 1994, 1995.

benefits) that many enterprises continue to provide for retirees and the replacement rate of the total compensation package comes close to 90 percent. In most countries the replacement rate is 40 to 60 percent (see table A2.3).

There is broad consensus in China that in the long run the replacement rate should gradually be brought down to about 60 percent of wages (net of payroll taxes). This would also create incentives for retirees to continue in some form of productive employment;

encourage workers to save in their own accounts, thus maintaining a high national savings rate; and keep some pressure on children to provide financial (and emotional) support to retired parents.

Indexation of pensions

The current approach (also emphasized in the draft Social Security Law) seems to provide for indexation of pensions to wages. This allows retirees to enjoy the benefits of current labor productivity growth and facilitates adjustment to the transformation of in-kind social benefits (housing, health care, and the like) to cash wages. In most cases, however, the indexation coefficient to wage growth is nontransparent. For the provinces and cities for which data were made available by the Chinese authorities, the coefficient ranges from 40 percent to 80 percent. This is, of course, a very uncertain form of indexation. For example, when nominal wages rose 30 percent in 1994 and the overall price index rose 23 percent, a 40 percent indexation to wage growth meant a reduction in real pensions of 11 percent. If wages had instead been growing at 25 percent a year and inflation were 8 percent a year, a 40 percent indexation would have meant that real pensions were rising 2 percent a year.

Indexing of pensions is a complicated matter. It could be argued that pensioners should share in the increasing prosperity of workers (although it may be hard to persuade pensioners to share in a decline in incomes of current workers). This would maintain horizontal equity, and pensioners would not feel deprived relative to current workers. It could as well be argued that pensions are implicit compensation for incomes withheld (implicitly or explicitly) during a pensioner's working life and thus that pensions related to real wages during that period are appropriate. Switzerland indexes pensions to the arithmetic mean between price and wage inflation, which is equivalent to linking pensions to 50 percent of the real wage growth.

In China's conditions, real wages are expected to rise fast (5 percent or more for ten years or longer), and indexation to real wages would substantially increase the pension burden (including transition costs). Considering the expected sharp rise in system dependency ratios by 2030 and the difficulties of finding finance for the transition, it seems advisable to lower the expectations of pensioners. If the funding situation improves, higher pensions in line with real wage increases can be granted later. At this stage a more affordable adjustment mechanism, in line with general international practice and the stated objectives in many localities, would be full indexation to the consumer price index.

Extending coverage of the pension system

Pension coverage is low in the nonstate sector. Farmers and township and village enterprise workers are covered only under voluntary pension schemes organized by the Ministry of Civil Affairs (see box 1.1). The system provides very little old age security. Many township enterprises are now located near cities and have fixed employment. Their workers, whose links with farming are fast disappearing, have all the characteristics of formal employment and should be covered in the formal pension system. Extending coverage to them would be desirable for their old age security and for the transition to a new pension system.

The problem is that these enterprises with low pension coverage are also the most dynamic. At a time when economic development and job creation are major concerns, any policy that would substantially increase labor costs for these dynamic enterprises must be pursued with caution. The high contribution rates in the state sector cannot simply be extended to the nonstate sector, including township enterprises. The key may be to set a low mandatory contribution rate, while allowing for the special needs of certain occupations (such as coal miners or railway workers) through a supplementary scheme.

Administrative responsibility needs to be rationalized as well. The Ministry of Civil Affairs, which is currently in charge of pension issues for township enterprises, may be reluctant to cede responsibility to another existing ministry. It might be easier to transfer responsibility to an entirely new national agency that would take over pension responsibilities from all ministries. Coverage would be extended to township enterprises only gradually, beginning, perhaps, with enterprises having more than fifty workers.

Establishing funded individual accounts

There is growing consensus in China (reportedly reinforced by strong policy directives from top economic management) that a substantial part of pensions (say half or more) should come from fully funded individual accounts. Individual accounts have clear advantages when the population is aging rapidly and economic growth prospects are uncertain (World Bank 1994a), although empirical evidence on these advantages is not robust across countries. Official Chinese forecasts indicate that the pay as you go system will require very high (38 percent) contributions by 2030, when growth rates may be slowing. Rates that high could lead to widespread evasion by enterprises and a high burden on society. Funded individual accounts would shift the burden of savings to current workers, who would be looking after their own retirement costs and not depending on their children's contribution. In the implementation phase, individual accounts create incentives for workers to ensure that enterprise make their contributions and that fund managers maximize the rate of return on pension funds.

However, individual accounts do not reap their full individual or social benefits if they are notional (see box 2.2). The link between individual contributions and individual benefits exists, but there is no opportunity to maximize the rate of return, which becomes adminis-

tratively determined. Nor is the objective of term transformation of savings achieved. And notional accounts cannot prevent the problems stemming from the demographic transition, since contribution rates will have to rise dramatically to pay for the promised pensions as the population ages.

Thus the preference for substantial funded individual accounts is appropriate. The system is not without risks, however, both to the individual and to society. Investment fund managers can make mistakes, investment companies can go bankrupt, and individuals may be left without much pension in the end. Also, people with low wages or irregular employment may not have much in their individual accounts at retirement. Society would need to provide protection in such cases, by guaranteeing a floor level of benefits below which society does not want any individual to fall.

Reforming the financial sector

Individual accounts cannot provide a reasonable level of old-age security unless interest rates are close to the growth rate of wages. In their calculations of replacement rates on personal accounts, Chinese agencies

---

BOX 2.2

What is a notional pension account?

Some cities in China, including Shanghai, are experimenting with notional accounts that are essentially pay as you go. In a notional system, a worker's account is set up merely as a bookkeeping device, to keep track of contributions plus imputed interest at a rate determined by the government. But funds are never actually accumulated in these accounts. When a worker retires the notional accumulation in the account is converted into an annuity which is paid to the retiree using the current contributions of younger workers.

Suppose that a worker about to retire has contributed 10,000 yuan and been credited with 2,000 yuan in interest. The 12,000 yuan would be converted into an annuity of approximately 100 yuan per month. When the worker retires, there are no assets in the account to finance the annuity. Instead, the pension is financed on a pay as you go basis by the contributions of younger workers, who are also accumulating notional accounts.

Notional accounts are attractive to countries that want to reform but that find a shift to full funding difficult to achieve because they already have a large public pension debt. A plan based on notional accounts accomplishes some but not all of the objectives of a reform. It produces a close transparent relation-

ship between contributions and benefits, thereby deterring evasion and other distortionary behavior. It eliminates some undesirable redistribution within the same cohort of individuals. It automatically adjusts retirement age up or benefits down as expected lifetimes lengthen, thereby keeping pension costs from rising as fast as they would otherwise.

But there are many objectives a notional account cannot achieve. It may not be portable when a worker changes jobs. It does not increase long-term national savings. It will continue to require a sharp rise in contribution rates as populations age, producing large intergenerational transfers. And interest rates on notional accounts are vulnerable to political manipulation, affecting future tax burdens in unexpected ways. Since notional accounts do not accumulate assets or generate investment earnings to cover the promised annuity, they leave governments with the full responsibility of covering the annuity on a pay as you go basis.

If China wishes to avoid sharply rising payroll taxes, it must avoid notional accounts and find a way to fund individual accounts, accumulating assets that can be used to build the productive capacity of the economy during a worker's active phase and pay pensions during the retirement phase.

assume the two to be equal. They also recognize that rate of return on pension funds can be high only if the general interest rate structure is rationalized—special rates on pension funds can be only occasional exceptions, not the general rule.

To Chinese officials the case for rationalization of interest rates and deregulation of the financial sector appears more persuasive from the perspective of pension funds and savers than from the perspective of investors. The link between low interest rates and loss of real value of pensions is arithmetically demonstrable and is more persuasive in terms of the pain reduced pensions can inflict on the elderly. Thus funded individual accounts could become a pressure point for rational interest rate policy.

The problem lies in the impact of higher interest rates on the financial position of state enterprises and the government. A simulation analysis shows that a 10 percentage point increase in interest rates would cause a net loss of 100 billion yuan for state enterprises and 88 billion yuan for the budget (World Bank 1995e). Raising interest rates on longer-maturity loans so that they could be used for longer-term investments is a possible short-run solution. Borrowers can afford to pay higher (at least 3 percent) real interest rates given the high productivity of capital in infrastructure and other long-term assets in China. We assume here that China will move in the medium term to market-determined interest rates that remunerate savings, which is crucial to the success of pension reforms.

Estimating transition costs

While there is a growing consensus on moving toward individual funded accounts, there is no consensus on how to finance the transition from pay as you go to fully funded individual accounts. With fully funded individual accounts, the current contributions of enterprises and workers would go to individual accounts and so would not be available for paying current pensions. Additional financial resources are needed to pay current pensions and the accrued pension rights of workers who had contributed before the start of individual accounts. The problem then is how to make these "double" payments?

The first step is to calculate the "implicit pension debt," the benefit promises made to workers and pen-

sioners. Although the implicit pension debt is not legally binding in many countries, there is usually a social and political obligation to provide workers with pension benefits in their old age. The implicit pension debt is the present value of the benefits that would have to be paid to current pensioners and that have been accrued by current workers. (See box 2.3 on how implicit pension debt is usually calculated and how other countries making the transition are paying the implicit debt.)

Some rough calculations of the implicit pension debt made by Chinese policy analysts suggest that the costs could be three to four times GDP. The size of the debt has apparently discouraged the authorities from recognizing the debt explicitly. They have instead continued to increase the contribution rates and extended the coverage of the pension system. By the government's own analysis average contribution rates by 2030 would be in the range of 28 to 39 percent. Rates that high are clearly unsustainable and would lead to widespread evasion and loss of viability for many nonstate enterprises. It is thus important to estimate transition costs carefully and to explore whether they can be paid in ways other than by taxing current workers.

The intended benefits level for pensioners and current workers, the age structure of the working population, and the life expectancy of the retirees together determine the size of the implicit pension debt. International experience suggests that the debt is in the range of 20 to 30 times pension payments (see box 2.3). China's pension payments in 1994 were about 2.3 percent of GDP, implying an implicit pension debt of 46 to 69 percent of GDP in 1994. These broad estimates are in line with the results of more detailed calculations based on explicit modeling of the benefits and age structure of the population in China (see table 3.9). Estimates for Shanghai and Shenyang, two cities with a serious aging problem in the state enterprise sector, showed an implicit pension debt of one-third to one-half of municipal GDP for Shenyang and about 100 percent of municipal GDP for Shanghai (World Bank 1996).

Thus it is reasonable to conclude that China's implicit pension debt for the enterprise sector is less than 50 percent of GDP, well below the 80 to 120 percent of GDP in Chile and Argentina when they made the transition from pay as you go to funded systems.

BOX 2.3

The implicit pension debt

The concept of *implicit pension debt* refers to the benefit promises that a pension scheme makes to workers and pensioners. Future pension entitlements are usually not included in assessments of the public sector's financial situation. But as aging populations put increasing pressure on pay as you go schemes, concern is growing that contribution rates might have to rise to unsustainable levels and that governments will have to assume an increasing share of pension expenditures in the future. The concept of *implicit pension debt* has been introduced to illustrate the magnitude of the pension burden and to compare pension promises across countries. The results depend strongly on many economic and demographic assumptions and are thus subject to much debate, but they do give an indication of the financial burden that economies will face in the future due to their pension schemes.

The size of the implicit pension debt (the present value of benefits that will have to be paid to current pensioners and current workers) depends on several important factors.

One is the coverage of the pension system. In countries with low coverage relative to the total labor force, the pension debt is a small portion of GDP. As coverage of a pay as you go scheme expands, the implicit pension debt also grows, because many additional workers will be promised future benefits. Another factor is the age distribution of the population. The more pensioners and workers there are who are close to retirement, the higher is the pension debt. A third factor is the level of benefits. Generous pension promises mean higher payments in the future. But since the debt is only implicit, benefit promises can be gradually cut back by raising the retirement age, reducing the wage replacement rate, and changing the indexation mechanism.

The discount rate chosen for present value calculations (discounting recognizes the fact that money received next year has less value than money received today) also affects the results. A high discount rate yields a lower present value of the pension debt. There is considerable debate on the choice of discount rate. Many calculations have used a discount rate of 4 percent, the current average long-term real interest rate on government bonds in major OECD countries. Some experts argue that the long-term interest rate is higher than that and that a rate closer to the returns of private pension funds should be used, especially if beneficiaries regard promised benefits as risk-bearing assets. A case can also be made for a discount rate lower than capital market returns, because of the public provision of annuities in social security schemes.

Calculating the unfunded liabilities of a pension system further requires a set of assumptions about economic and population growth, wage growth, vesting rules, and the future rules of the pension system. For countries without adequate records on the contribution history of workers, the average pension for new pensioners can be used as the basis for calculating workers' accrued rights.

The pension debt is most commonly calculated on the assumption that the unfunded system would be terminated immediately and that all pensioners and workers would have to be compensated for their future pensions and accrued rights (the *termination hypothesis*). It does not take account of possible new obligations or income from future contributions or interest. Other calculation methods assume that the system is closed to new entrants but continues in effect until the last current contributor dies (future contributions from and benefits to all current contributors are thus counted) or assume an open system by estimating the present value of all future pension payments including those to new entrants.

Under the termination hypothesis, the pension debt ranges from 100 to 200 percent of GDP in most OECD countries. In Hungary and Uruguay, which have pension systems with high coverage, high system dependency ratios (relatively few workers to support a growing number of pensioners), and generous benefits, pension debt is more than 200 percent of GDP. Estimates of the Chilean pension debt at the time of reform range between 40 and 130 percent of GDP, depending on the discount rate used. Estimates for Peru put pension debt at 40 percent of GDP and for Colombia at 90 percent when they moved from unfunded to funded pension schemes.

Fully funded schemes have no implicit pension debt since each generation saves for its own retirement. Funded schemes accumulate assets from the start, sufficient to cover future pension liabilities. Changing from a pay as you go to a funded pension system makes the implicit debt explicit. Workers' contributions can no longer be used to pay for current pensions, so the "debt" owed to current pensioners has to be financed from a different source, unless the government wants workers to pay a double burden. Still, the financing of the implicit debt can be stretched out over a long period if desired.

In China, the low implicit pension debt should make it relatively easy to reform the system and to finance the benefit obligations to pensioners and workers nearing retirement. If the system is not reformed soon, however, the pension debt will grow rapidly. As coverage expands and workers age, it will become much more difficult to move from the pay as you go to a more funded system. With total pension expenditure amounting to only about 2 percent of GDP, China is in a very good position to move to a partially or fully funded pension system.

Financing the transition

Countries have used a mix of instruments to pay off the pension debt and bridge the financing gap when they changed from a pay as you go to a funded pension scheme. The financial instruments available are the same as those used to finance any other public expenditure: bonds, higher taxes, lower spending, and transfers of public assets. The mix of instruments chosen will determine the impact of pension reform on economic growth and the distribution of the costs and benefits of transition between and within generations.

Experiments with different sources of financing for pension expenditures are already under way in China, although they are meant to provide additional resources for pension pools rather than to finance the transition gap. In Luoyuan county in the province of Fujian part of the tax on land-use transactions is transferred to the pension pools. In Zhengzhou City, pension pools receive part of the proceeds from the sale of shares of corporate state enterprises. In Shantou in the province of Guangdong enterprises pay a tax of 1.6 yuan per thousand in retail sales or 2.7 yuan per thousand in wholesale revenues in addition to the contribution rate to the pension pool. In Tongling City some shares of previously state-owned enterprises were given directly to the pension pool and some were given to workers individually. In Nanyang City poorly performing enterprises pledged a proportion of the revenue from the lease of fixed assets, equipment, factory, and warehouse space for the payment of pensions and the redeployment of workers.

Beyond these specific measures the government may decide, as in Chile, to cover the financing gap with government bonds, exchanging new explicit debt for old implicit debt. If the government issues bonds, future generations will have to pay higher taxes to repay them. Debt financing is thus similar to the pay as you go system except that it allows the financial burden to be spread out over a longer time.

Much of the demand for new government bonds would come from the pension system itself if it is funded, especially in China where the financial sector is still underdeveloped. In Chile the new pension funds were allowed to invest only in government bonds and bank deposits in the first years after reform. As the financial sector developed and more instruments became available, pension funds were allowed to diversify their investments. Debt financing can pave the way for pension funds to contribute to financial sector development and to play an increasingly important role as institutional investors. Chile also used a budget surplus to cover part of the transition costs. China has some budget surpluses in its current account but has deficits in the overall account of about 2 percent of GDP. Since the government intends to further reduce its borrowing program, it is unlikely that the budget can provide much funding for transition.

Another approach is based on the recognition that the implicit pension debt went into the creation of enterprise assets, so that those assets should be used to redeem that debt. This principle is now recognized by the Chinese government in regulations that provide for the proceeds from the sale of assets of a liquidated enterprise to be first used to meet pension obligations. Some municipalities have been considering such uses of assets for solving the pension debt problem. The government has issued instructions prohibiting the use of state assets for pension funds. Why it did so is not known. One possibility is that the government is concerned about state asset-stripping and the siphoning off of state assets for pension funds, leaving behind inadequate assets for meeting other obligations and for the smooth functioning of enterprises.

The assets and liabilities of state enterprises should be evaluated carefully and systematically. The process has already started, and the results should be available soon. Preliminary assessments in 1995 suggested productive assets of more than 2 trillion yuan and "unproductive" assets such as land and housing of more than 5 trillion yuan. The explicit debt of state enterprises to banks is less than 2 trillion yuan, while pension debt and unemployment debt should be less than 2 trillion yuan.[1] Overall, then, state enterprises are in a solvent position. Moreover, since the state owns the assets and owes the liabilities, an asset-liability swap will be easier now than later, when multiple ownership of assets and liabilities will develop.

While keeping open options for funding transition costs, special attention should be focused on how state enterprise assets could be used to help in the transition from a pay as you go system to funded individual accounts. Two principles apply:
• Any asset transferred to pension funds has to be marketable and to carry a value corresponding to the market-determined price. The scheme should not require cashing in a large proportion of assets in the short run. Rather, enterprise assets should be sold gradually.
• The link between asset sales and beneficiaries should be as close as possible. If enterprises are seen as ultimately responsible for pensions, the first claim on their assets should be for settling their pension debt. If assets

and pension liabilities are located in the same place, the swap would be easier to understand and to implement.

China's advantages in bearing the transition costs

Not only are China's costs of transition to a funded system lower than those of many other economies, but its financial capacity to fund the transition is also probably greater, for at least four reasons: benefits from unification of the pension system, rapid growth of income, structural change in the economy, and capital gains in state assets. These facilitating factors will be present over the next fifteen to twenty years, but they will largely disappear after 2030. They cannot be counted on to finance high pension costs indefinitely, but they are ideally suited to finance the temporary needs of the pension system transition.

• *Unification.* If the fragmented pension system is unified and coverage is extended to the nonstate sector (including township enterprises), this would generate substantial resources to fund the transition. In expanding coverage, care would have to be taken to keep contribution rates low so as not to discourage economic growth.

• *Rapid growth.* China's economy is growing at more than 10 percent a year, and its domestic savings rate is more than 40 percent of GDP. Rapid growth increases the size of the economic "pie" and makes it feasible to use a slice of the incremental income to fund the transition. An important mechanism for accomplishing this is to use price indexation rather than wage indexation of benefits, so that pension costs do not rise commensurately with growth and so the growth dividend can be siphoned off to pay for pension reform.

• *Structural change.* Associated with rapid growth of GDP is the rapid structural change of the economy, as labor shifts from rural to urban areas, from agriculture to nonagriculture activities within rural areas, and from the state sector to the more efficient nonstate sector in urban areas. The reserve pool of surplus labor will enable the nonagricultural labor force to expand rapidly, even if the overall labor force grows slowly, thereby expanding the contributory base for the pension system. If the transition has been made by the time these workers retire, their pension needs will be met, in part, through their own funded individual accounts and the higher productive capacity generated by their retirement savings.

• *Capital gains.* The state enterprise sector has not collapsed in China as it did in other socialist economies in transition. It provides about 70 percent of government revenues and accounts for a significant part of the country's impressive performance in savings, exports, and GDP growth. The value of assets in the state enterprise sector is several times GDP; by comparison, it is estimated to be less than 50 percent of GDP in Latin American and Eastern European countries. The availability of these assets opens up possibilities for transition financing in China that are not available in other economies.

• • •

The discussion thus far has identified various components of reform in China's pension system that seem advisable, as well as the broader macroeconomic factors that make these reforms financially viable. There are, however, risks and uncertainties associated with the institutional and political feasibility of adopting various components of these reforms, as there are with the macroeconomic environment. The following chapter therefore considers the *quantitative* implications of various packages of reform and macroeconomic assumptions.

Note

1. Unemployment debt is defined as the payments that would have to be made to workers who were laid off. For estimation of this debt, see Guo (1995).

chapter three

Quantitative Analysis for a Preferred Pension System

Experimentation in design and gradualism in implementation have been the hallmark of China's successful transition from a planned to a market economy. In the pension system, however, experiments have been going on for nearly ten years. The new plans, whatever they are, should be put in place by 2000 in order to end the uncertainty and to take advantage of the special opportunities for reform that exist today. What follows is a set of recommendations on a master plan for a pension system. Some specific numbers are postulated for illustrative purposes, but actual calculations would have to be based on local circumstances. Implementation will no doubt proceed gradually, with experiments in certain localities. Though suggestions are made here for a preferred pension system in China, it is fully recognized that there is no unique solution to China's pension problem and that final decisions have to be made in light of social and political considerations that are not discussed here.

Demographic and macroeconomic developments

The financial viability of the pension system depends crucially on such economic variables as growth of the labor force, wages, and interest rates. Making some basic assumptions about these variables is essential for assessing the financial outlook of the pension system. While it is beyond the scope here to make detailed investigations of these macroeconomic variables for projections, some illustrative projections are presented. The model developed for this study is capable of generating quick responses to alternative sets of assumptions and is intended to be an interactive tool for discussions with policymakers.

Population projections

Projections for China over the period 1995–2050 are for no growth in the working-age population between 2020 and 2030, followed by a decline after 2030 (table 3.1). Meanwhile, the population in the age group 65 and older will rise steadily and rapidly until 2050. As a result, the demographic dependency ratio goes from 8.7 percent in 1990 to 31.2 percent in 2050 (figure 3.1).

Labor force projections

While the demographic dependency ratio shows the economywide need for supporting the elderly, what matters for the financial viability of the pension system is the system dependency ratio, the ratio of the elderly covered under the pension system to the number of workers in the system. The key baseline assumptions used to obtain the projected system dependency ratios are as follows (table 3.2):

• *Participation rate.* The labor force participation rate depends on the legal retirement age and a host of other socioeconomic factors. Here we assume that the legal retirement age will be gradually increased to sixty-five for men and women. While this change will substantially increase the overall participation rate, the participation rate in the age group 15–20 may decline as the economy develops and people stay in school longer. It is assumed that the participation rate—already high by international standards—will increase slightly from about 77 percent in 1994 to 80 percent.

• *Declining share of employment in agriculture.* The 2.5 percent annual rate of decline in the share of the labor force in agriculture between 1990 and 1994 is projected to continue, so that the share of labor in agriculture falls to 17 percent by 2030 and 10.5 percent by 2050. These projections are consistent with shares for the Republic of Korea (15 percent in 1994) and Japan (7 percent in 1991).

• *Urban-rural distribution.* Between 1990 and 1994, employment in urban areas grew 3.4 percent a year, while that in township and village enterprises grew 6.9 percent. Township enterprises are beginning to face some difficulties, however, and employment declined in 1994. But because government policy strongly supports the development of small and medium-size towns, where township enterprises are the main sources of employment, growth rates of employment in township enterprises are assumed to converge to the same rates as in urban areas. The nonagricultural employment distribution will stabilize at 60:40 between urban and township enterprises. The government's policy of avoiding excessive population concentrations in cities will keep the share of urban employment in China at around 50 percent of the total labor force, significantly below the 70–80 percent reached in countries such as Japan, Korea, and the United States.

• *Share of state and nonstate sectors.* Employment in the state sector (government organization, public institutions, state enterprises, and collectively owned enterprises) has grown only marginally (1 percent a year), from 139.0 mil-

---

TABLE 3.1

Projected population of working-age group and the elderly

Indicator	1990	1995	2000	2010	2020	2030	2050
Population in age group 15–64 (millions)	762.0	808.3	845.8	955.9	988.6	989.4	962.2
Population in age group 65 and older (millions)	66.1	75.9	86.6	104.2	153.6	214.9	300.4
Demographic dependency ratio (percent)[a]	8.7	9.4	10.2	10.9	15.5	21.7	31.2

a. Ratio of 65 and older age group to 15–64 age group.
Source: World Bank 1994b.

China's population is aging fast

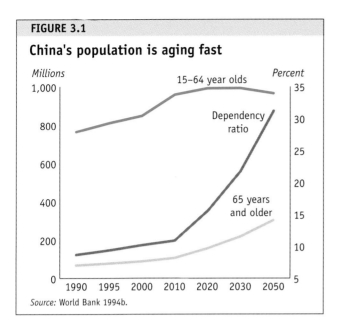

Source: World Bank 1994b.

lion in 1990 to 145.0 million in 1994. Recent government policy statements suggest a reduced role for government over time, and employment in the state sector may also shrink. Thus employment in the state sector is assumed to stabilize at the 1994 level, and the share of the state sector in nonagricultural employment is projected to decline from about 52 percent in 1994 to 21 percent by 2050.

Labor productivity, real wages, and GDP

Between 1990 and 1994 the average productivity of labor (GDP in nonagricultural sector divided by the

labor force in nonagricultural sector) grew at 8.9 percent a year, while real wages grew at 6.4 percent. The rapid growth in productivity was associated with a high level of capital formation and broad opportunities for technological catch-up with more advanced countries. A continued high level of investment in China is both feasible and desirable over the medium term, and opportunities for technological catch-up will remain significant for some time. Therefore, the growth rate of labor productivity and GDP should remain high up to 2010.

Over the longer term, however, these growth rates will come down substantially, as savings rates decline with rising dependency ratios and the scope for technological catch-up diminishes. It is assumed that average real wages will grow at 4 percent a year during 1995–2020, decline to 3 percent during 2021–30 and to 2 percent thereafter. (For individual workers, a merit increase of 1 percent per year of service is assumed in addition to the average wage increase.) It is assumed that the real rate of return on funds in the pension system will equal the growth rate in real wages, including merit increases.

The projections clearly suggest that by the time a young worker who is twenty today retires around 2040, China would have many of the characteristics of the old age crisis that mature OECD countries now have. The system dependency ratios will have risen to 50 percent, while the growth rates of the labor force, labor productivity, and GDP would have slowed considerably.

TABLE 3.2

Labor force projections

Year	Population in age group 15–64 (millions)	Participation rate (percent)	Share of labor in agriculture (percent)	Urban employment (millions)	Employment outside public sector[a] (millions)	Employment in township enterprises (millions)
Actual						
1990	762.0	74.5	60.0	147.3	8.3	92.0
1994	799.2	76.9	54.3	168.1	27.1	120.0
Projected						
1995	808.3	77.0	52.3	172.8	27.8	115.2
2000	845.9	80.0	42.7	224.4	79.5	149.6
2010	955.9	80.0	29.1	315.5	170.5	210.3
2020	988.6	80.0	21.9	359.7	214.7	239.8
2030	989.4	80.0	17.0	382.6	237.6	255.0
2040	950.2	80.0	13.7	381.8	236.8	254.5
2050	962.2	80.0	10.5	400.9	256.0	267.3

a. Defined as urban employment minus employment in state-owned units and collectively owned enterprises. A 3 percent unemployment rate is incorporated in the projection.
Source: World Bank 1996 (tables TA4.1–4.3).

Unless new retirement systems are set up now for current young workers, they will face significant pension uncertainty in their old age.

Simulation results with limited reforms

This section presents some quantitative implications for a few of the many possible assumptions on demographic and macroeconomic variables and pension system policy variables, using the pension system model designed for this study.[1] The results reported here are intended only to illustrate how the model could be used to assist in policy discussions on pension system reforms.

On the benefit side of the pension system the key indicator is the replacement rate as a percentage of net wages. On the cost (and affordability) side the key indicators are the system dependency ratio, sustainable contribution rate, and accumulated reserves in the pension system.[2] The sustainable contribution rate differs from the balanced contribution rate in that it attaches more importance to financial viability over a very long term. (For details on sustainable contribution rates in industrial countries, see IMF 1996a.) The results of simulations with limited reforms are described below and presented in table 3.3 and figure 3.2.

In *scenario 0* the pension system retains many of its current characteristics. It covers mainly the state sector with just 10 percent coverage of the nonstate sector and no coverage of township and village enterprises. Pensions are indexed to the average increase in prices and nominal wages. The wage replacement rate is 80 percent, and the retirement age is 55 for women and 60 for men. Compliance is 85 percent, and real wages and interest rates are projected to grow at 5 percent a year during 2000–10, 4 percent during 2011–30, and 3 percent thereafter.

The simulation results confirm the views of Chinese analysts that the current system is not viable in the long term. The system dependency ratio rises to 76 percent by 2050 and the sustainable contribution rate to 45.9 percent (baseline scenario in figure 3.2).

In *scenario 1* the situation improves somewhat because pensions are not indexed to real wages. However, the sustainable contribution rate is still too high at 40.8 percent.

In *scenario 2* improvements are greater because the coverage rate is increased to 50 percent by 2010 for both township enterprises and the nonstate urban sector. The sustainable contribution rate declines to 37.4 percent.

In *scenario 3* the replacement rate is reduced to 60 percent and the sustainable contribution rate falls substantially to 28.4 percent.

Scenario 4 includes all the reforms already mentioned plus an increase in retirement age to sixty-five by 2040. As a result of this change the number of retirees goes down and the number of workers goes up, and the situation changes dramatically. The system dependency ratio plunges from 21 percent in 1995 to 15 percent in 2010 and then rises to 47 percent in 2050. The sustainable contribution rate goes down to 19.7 percent, which is moderate relative to the average contribution rate of more than 20 percent for most enterprises in China today (baseline scenario with nonstructural reforms in figure 3.2). The system now looks viable except for the risks noted below.

The defined benefit system makes long-term commitments to pay benefits while its receipts are subject to the uncertainties associated with future macroeconomic and demographic developments. If unemployment rises, pension system receipts may fall without a corresponding decline in pension payments. And as the pension system expands to cover the nonstate sector, the compliance rate may drop. Further, rates of return on pension funds may suffer if they are managed by the government and become vulnerable to political pressures.

A simulation of these downside risks illustrates how the system can become nonviable again. The participation rate goes down from 80 percent to 70 percent, the compliance rate falls from 85 percent to 60 percent, real wages rise faster, and the rate of return on pension funds goes down (see table 3.3, scenarios 5–9). The system dependency ratio rises to 50 percent by 2050, and the sustainable contribution rate rises to 30.4 percent (the downside scenario with nonstructural reforms in figure 3.2).

Thus reforms are needed to improve the financial viability of the system. The reforms should aim first at sharing risks among society, individuals, and enterprises and second at reducing risks in the system. One way to do both is to move to a multipillar system in which society bears the risks for a small defined benefit component and individuals bear the risks for a larger

TABLE 3.3

Simulations with nonstructural reforms
(percent)

	Scenario 0	Scenario 1	Scenario 2	Scenario 3	Scenario 4	Scenario 5	Scenario 6	Scenario 7	Scenario 8	Scenario 9
Sustainable contribution rate[a]	45.9	40.8	37.4	28.4	19.7	20.4	23.1	22.9	24.3	30.4
System dependency ratio										
1995	26	26	21	21	21	21	21	21	21	21
2000	27	27	19	19	18	18	18	18	18	18
2010	34	34	17	17	15	15	15	15	15	15
2020	49	49	27	27	18	19	19	19	19	19
2030	73	73	57	57	29	31	31	31	31	31
2040	78	78	78	78	50	53	53	53	53	53
2050	76	76	77	77	47	50	50	50	50	50

System characteristics (changes effective in 1995)

	Scenario 0	Scenarios 1–9
Real wage indexation	50	Scenario 1: scenario 0 with no wage indexation
		Scenario 2: scenario 1 with increase in coverage rate to 50% (by 2010)
Coverage rate		Scenario 3: scenario 2 with decrease in replacement rate to 60% for retirement and 30% for disability
Township enterprises	0	
Other enterprises	10	Scenario 4: scenario 3 with increase in retirement age for men and women to 65 (by 2040)
Benefits (replacement rate)		Scenario 5 : is scenario 4 with decrease in participation rate to 70%
Retirement	80	Scenario 6: scenario 5 with decrease in compliance rate to 75%
Disability	40	Scenario 7: scenario 6 with changes in wage increase rate: 6%, 5% (2011), 4% (2031)
Survivor (lump sum)	100	
		Scenario 8: scenario 7 with changes in interest rate: 4%, 3% (2011), 2% (2031)
Retirement age		
Women (years)	55	Scenario 9: scenario 7 with changes in interest rate: 3%, 3% (2011), 2% (2031) and decrease in compliance rate to 60%
Men (years)	60	
Participation rate	80	
Compliance rate	85	
Wage increase rate (including seniority increase)	5 (2000–10) 4 (2011–30) 3 (2031–)	
Interest rate	5 (2000–10) 4 (2011–30) 3 (2031–)	

a. Rate at which accumulated reserves in 2050 are ten times the annual deficit in that year.
Source: World Bank 1996 (technical annex 4).

defined contribution component. The individual account system helps reduce risks relating to compliance rates and rates of return since individual workers become powerful allies in improving compliance rates.

Individuals keep track of the contributions made by enterprises into individual accounts, helping to ensure that enterprises are making proper payments to the public pillar. Concerned about maximizing the rate of return on savings through the pension system, individuals also become a pressure group for achieving reforms in the financial sector and in pension fund management. The performance of independent pension fund companies, in turn, becomes a yardstick for measuring the performance of the pension fund administration's management of the public pension pillar (see chapter 4).

The simulation results on the financial viability of the pension system (including the risk factors) thus validate the general direction of reforms proposed by Chinese policymakers, including the switch to a multipillar system with individual accounts. However, the viability of these reforms will depend on how pension responsibilities are divided among different agents and what arrangements are made for funding the transition. It is to these issues that the report now turns.

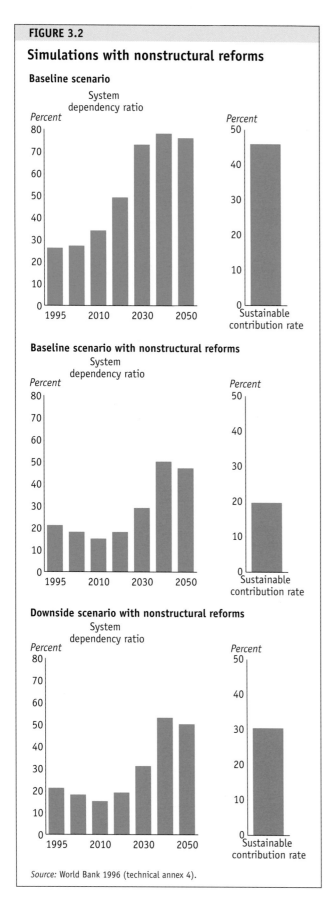

Source: World Bank 1996 (technical annex 4).

A conceptual framework

This section reviews the quantitative implications of a proposed reform package. Before going into specific numbers, however, it is helpful to summarize the conceptual rationale for the proposed system, as it emerges from the discussion so far.

The extensive social welfare systems in OECD countries have become an economic and in some cases a political liability. Many of these countries are trying to reform their social welfare systems and to reduce the role of the state. Developing countries like China should learn from the mistakes of the industrial countries, rather than repeat them. The obvious question then is, why should China develop an extensive pension system that may end up burdening the state and accelerating the break up of the traditional system of old age security based on the extended family? When the global trend is toward reducing the role of the state, why not leave old age security to private provision?

The short answer is market failure (box 3.1). People may not save enough when they are young, because of shortsightedness. There may not be enough instruments of long-term savings to protect savings against inflation, and private insurance markets may suffer from adverse selection (see glossary). As the economy grows and workers follow jobs to new locations, the family-based system of old age security is likely to weaken. By default, society will be obliged to take care of the elderly, particularly salaried workers who, unlike farmers and the self-employed, may have no income-earning assets. Better for the state to anticipate the problem and design a system now, before it becomes a crisis. At the same time, of course, China should learn from the mistakes of the OECD countries and design a system that minimizes the problems that have emerged in OECD countries and that fits into China's realities.

One important choice is between defined benefit and defined contribution systems. Most OECD countries opted for a defined benefit public pension system. It has the advantage of an assured income for retirees but the problem of uncertain financial health for the agencies providing the benefits. Pension agencies face a mismatch between defined long-term obligations and undefined long-term receipts, which are dependent on macroeconomic factors beyond the agencies' control or knowledge. The problems are intensified by the reduced incentive for compliance when benefits are not linked to contributions. In a defined contribution system, uncertainties are shifted to the individual but the incentives for compliance and saving remain strong.

In China, as in many other countries, it seems advisable to combine the two approaches. A defined benefit system would provide the basic minimum poverty-level income, while a defined contribution system would be geared to meeting the needs of the elderly according to their social and occupational position. All publicly supported benefits should be low enough to keep some pressure on the elderly to seek gainful employment after retirement, and some pressure on the extended family to provide economic (and emotional) support to the elderly. And because China's informal sector (including agriculture and the self-employed) will remain a dominant sector for some time, the lessons of the formal sector pension system could guide the extension of the system as time goes by.

BOX 3.1

Why should governments get involved? Failures of the marketplace

When traditional family-based arrangements for production and subsistence break down, formal market-based arrangements often replace them. Why do pure market solutions to the problems of old age security often fail? Why don't people simply save or purchase annuities when they are young, so that they can maintain a decent standard of living when they are old and less productive? Why do governments throughout the industrial world and increasingly in developing countries intervene in this area?

Market solutions—such as individuals saving and investing for their old age—fill some of the gaps left by the breakdown of the family system, but they fail to do the job completely. In the simplest case people may not save enough when they are young because they are shortsighted. Income security in old age requires very long-term planning, and many people lack the necessary information (about future health, cost of living, lifetime earning capacity, and the safety and productivity of alternative forms of investment and insurance).

Probably every society has some myopic members, though it is hard to prove that people are generally myopic—the evidence points in both directions. Shortsighted behavior becomes a social problem for two reasons. First, as people age they may change their preferences and wish that they had saved more—but by then it is too late. Second, if people do not save enough for their old age, the rest of society may feel obliged to support them. (People may take advantage of this sense of obligation by intentionally failing to save, a problem known as free riding and moral hazard.) Insufficient savings may also starve the economy of investable funds. So government policies that encourage or require people to save can make everyone better off.

But even if people try to save when they are young, they may find few reliable savings instruments, particularly in developing countries. They may not invest wisely, finding themselves with much lower returns than expected. Or investment returns may be low or negative for the entire economy during long spells, such as during the Great Depression of the 1930s.

Another set of problems stems from the absence of insurance markets. Since people are always uncertain about how long they will live, they may wish to purchase insurance that will pay them an income—a pension or an annuity—over their lifetime. But insurance companies are not well developed in many countries because of informational deficiencies and weak capital markets. When people have more information about their life expectancy than their insurance company does, a problem known as adverse selection occurs. Good and bad risks are pooled and premiums are charged according to the average risk of the group. The good risks (people who expect to die young) find these terms unacceptable. So the insurance company, left with only the bad risks (people who expect to live long), raises its premiums, leading more good risks to opt out. In the end, prohibitive prices put annuities beyond the reach of most people.

Even when annuities are available at a fair price, consumers must commit large sums of money irreversibly before age sixty-five in return for the promise of a lifetime pension paid out over twenty to thirty years. Many people would doubt the credibility of such long-term promises by private insurers. Private insurance companies also have a difficult time insuring against risks that hit everyone simultaneously, such as unexpectedly high inflation. For all these reasons, private annuity markets are undeveloped in most countries, and annuities that fully insure against inflation and recession are generally not available.

Thus when systems of communal living in extended families break down and before remedial government polices are firmly established, old people may have a hard time maintaining their standard of living because of saving or insurance problems. Furthermore, pockets of severe poverty have developed among those whose lifetime incomes were too low to cover minimally adequate consumption levels during their retirement as well as their working years (the long-term poverty problem).

It is clear, then, that governments should get involved when informal old age security arrangements no longer work. And the need for government involvement becomes more urgent as a country's population ages.

Source: World Bank 1994a (p.36).

Since the formal pension system will, for some time, cover only the formal sector, and largely in urban areas (which are already privileged in many ways), the state's role in the pension system should be carefully considered. The pension system should be seen largely as an insurance system (both social and private) rather than as a security system financed by general taxation or borrowing. Similarly, receipts and payments of the pension system should not become part of the regular fiscal budget. In order to pay for the externalities of the system (due to market failure, as noted above), the government could, as many others do, provide tax exemptions for contributions to the pension system and for incomes of the pension funds. For the defined contribution component the state may have to guarantee some minimum rate of return on pension funds. In exchange, it should closely monitor how the pension funds are used to ensure that the "public goods" character of the pension system is indeed satisfied and that the pension system does not become a tax shelter for the rich.

Three pillars

Taking all these factors into account, the best solution for China is a system that combines social pooling and fully funded individual accounts.[3] The proposed system is closer to the government's plan IIB than to plan I (see chapter 1) in that it provides a basic pension component to prevent retirees from falling below the poverty line. As noted above, such social protection is considered necessary for the elderly in the formal sector, who do not have the cushion provided by land (farmers) or physical assets (the self-employed). The proposal also has a substantial individual account component.

Pillar 1

The first pillar would be established on the defined benefit principle and provide poverty-level income to retirees with forty years of covered service. Whether extended on a universal or a means-tested basis, this pillar could be financed through general taxes or payroll taxes. The universalist approach has the advantage of administrative ease but the disadvantage of paying retirees who may be comfortably above the poverty line on their own. Means testing, however, is administratively complicated and attaches the stigma of penury to those who apply (an

important factor in a "face" conscious society). Financing out of general taxes causes less distortion of labor markets than payroll taxes but means that some redistribution will take place from uncovered sectors (agriculture and other informal sectors) to the formal sector.

In China's case it is recommended that this pillar provide universal benefits, financed out of payroll contributions, and covering all formal sector workers in urban areas and employees in large township enterprises. Pension payments will be about 24 percent of average wages (close to the urban poverty line in 1994). Over forty years of employment, this means providing 0.6 percent of average wages per year of covered service. Because the wage replacement rate would be higher for a low-income worker than for a high-income worker, the basic pension would help equalize the incomes of the elderly. This basic pension would be financed by enterprise contributions as a percentage of the wage bill. The exact percentage required will depend on a host of macroeconomic and pension system design considerations, as discussed below. For newly covered workers in the nonstate sectors, the basic payment could be 1 percent of the average social wage (average wage for urban workers) for each year of service, with a cap of 24 percent. This pillar would also provide disability benefits and survivor benefits.

Pillar 2

The second pillar would consist of mandatory individual accounts that are fully funded, with a close connection between contributions and benefits. These schemes could be employer-based or pure individual accounts. There are arguments on both sides. Individual accounts allow for individual choice and promote competition. These advantages depend critically on having well-informed consumers, which is not always the case. Administrative costs can be high, and individuals bear the entire risk of pension fund performance. Employer-based schemes may enjoy administrative economies of scale, and employers may bear part of the risks of pension fund performance. The employer may have too much control over the pension funds, however, and may borrow from these funds to the detriment of pensioners' interests. Moreover, employer-based pension schemes may impede labor mobility.

In China's case, where the enterprise sector is going through a major restructuring in which many enterprises may eventually shut down, labor mobility is crucially important. It is therefore advisable for the second pillar to be a fully funded individual account, financed 50:50 by workers and enterprises, with workers' contributions possibly increased over time as their incomes increase. In light of considerations noted below, it is proposed that the contribution rate be 8 percent of a worker's individual wage, which would generate a replacement rate of about 36 percent of final net wages after forty years of service, provided the rate of return on pension funds equals the rate of growth of wages. The government will bear part of the performance risk of pension funds by guaranteeing a minimum real rate of return on these funds. Survivor benefits and disability benefits will consist of the return of total contributions plus interest.

Pillar 3

The third pillar would consist of supplementary pensions based on voluntary contributions—by employers, individual workers in the formal sector, farmers, and informal sector workers. It would be part of the formal pension system to the extent that contributors claim tax advantages (up to a specified maximum) and channel their contributions to publicly licensed pension funds that are subject to government surveillance.

A natural question is whether this voluntary part of savings should be part of the pension system design at all. In China's case this seems advisable for two reasons. First, it would provide flexibility to the formal pension system by taking into account the special needs of particular localities and occupations, without which a unified system may not be acceptable. For example, for certain occupations such as coal mining, a higher replacement rate may be needed because of the particular difficulties of coal miners in supplementing their incomes after retirement; similarly, civil servants and other employees of public institutions expect a higher replacement rate in compensation for their relatively lower wages and salaries. Second, this pillar would provide an option for farmers and informal sector worker to become integrated into the public pension system at a pace that they consider appropriate. These workers may contribute to their individual accounts in licensed pension fund management companies or purchase life insurance policies and enjoy tax benefits up to a maximum allowable level.

Other voluntary arrangements

Individual workers may also supplement their pensions through voluntary savings and investment and through gainful employment after retirement and transfers within the family. However, these benefits would be in the nature of private goods and would be treated like any other income, without any special tax benefits. The public goods needs of the pension system would be served by the mandatory pension plans, which would provide wage replacement rates of above 50 percent for most workers.

Transition plan

To ease the financial costs of transition, it is proposed that the transitional benefits of current pensioners and current workers be held to a politically acceptable minimum. Current workers who will retire after the reform begins may receive the basic benefit, an accrual rate of 1 percent of salary for each year of service prior to reform (that is, the lower bound of the current proposals), and an annuity based on mandatory and supplementary contributions to their individual accounts after reforms.[4] Current pensioners will continue to receive their current pensions with full adjustment for increases in the consumer price index.

The proposed reforms are intended to establish a nationwide system of pensions (with local variations) that provide old age security to all covered workers. Two issues, in particular, need to be explored:
• What level of pensions would be available to workers at various income levels and age groups?
• Will the pension funds be financially sound enough to honor the obligations to old and new retirees?

Effects on individual pensioners

To illustrate how the proposed new system would affect individual workers, ten broadly representative cases are considered (table 3.4). The rate of growth in real wages (including a seniority-based increase of 1 percentage point a year) is assumed to be the same as the real return on pension funds: 5 percent a year for 1995–2010, 4 percent a year for 2011–30, and 3 percent a year thereafter. For the average-income worker—new and old—the replacement rate under the proposed system will be around 60 to 70 percent.

Case 1 is a male worker starting work in 1994 at the age of 20. His average wage is 3,000 yuan a year, 40 percent lower than the average social wage (5,000 yuan a year). When he retires at 65, his wage in 1994 prices will be about 18,700 yuan a year.[5] Pillar 1 will provide him with 40 percent of his final wage and pillar 2 with 35 percent, for a total replacement rate of 75 percent.

Case 2 is a female worker with the same age and wage characteristics as the male worker of case 1. However, she retires at 60. Her replacement rate is only 66 percent because she has only 40 years of contributions (as against 45 years for her male counterpart) and at age 60 has a longer life expectancy at retirement than the male worker at age 65. Thus her individual account is spread over a much larger number of years, leading to a smaller annuity. (If gender-specific life expectancy is used, the female worker's annuity would be even lower.)

Cases 3 and 4 illustrate the difference made by the relative income level of workers. Case 3 earns the average wage and has a replacement rate of 59 percent. Case 4 has a wage 40 percent above the average wage and has a replacement rate of 51 percent.

Cases 5 through 10 illustrate the situation of workers who have already worked for 20 or more years when the new system starts. For the male worker at age 55 with an average income, the replacement rate is 71 percent. For the male worker at age 58 with a wage twice the average social wage the replacement rate is 59 percent. As before, replacement rates for female workers are lower than those of their male counterparts. The lower pension benefits for female workers under the

TABLE 3.4

Pension and wages for ten hypothetical urban workers

Characteristics of workers

Case	Gender	Age in 1994	Annual wage in 1994 yuan	Year starts work	Retirement age	Year of retirement	Years of working life	Years of contribution
1	Male	20	3,000	1994	65	2039	45	45
2	Female	20	3,000	1994	60	2034	40	40
3	Male	20	5,000	1994	65	2039	45	45
4	Male	20	7,000	1994	65	2039	45	45
5	Male	40	6,500	1974	65	2019	45	25
6	Female	45	8,000	1969	60	2009	40	15
7	Male	50	3,500	1964	65	2009	45	15
8	Female	50	3,500	1964	60	2004	40	10
9	Male	55	5,000	1959	65	2004	45	10
10	Male	58	10,000	1956	65	2001	45	7

Pension receipts

Case	Gender	Annual wages at retirement (yuan)	Total annual pension (yuan)	Total pension as percent of net wage	Pillar 1 pension as percent of final wage	Total pension as percent of social wage	Past service contribution (yuan)	Accumulated contribution (yuan)	Annuity at retirement
1	Male	18,722	13,520	75.2	40.0	43.3	0	68,785	11.4
2	Female	16,150	10,297	66.4	40.0	38.3	0	52,778	13.8
3	Male	31,203	17,540	58.6	24.0	56.2	0	114,641	11.4
4	Male	43,685	21,561	51.4	17.1	69.1	0	160,498	11.4
5	Male	20,195	11,914	61.5	18.5	76.7	40,243	41,316	10.0
6	Female	16,631	8,429	52.8	15.0	81.1	47,872	20,451	11.5
7	Male	7,276	5,627	80.6	34.3	54.1	20,570	8,947	9.4
8	Female	5,701	4,093	74.8	34.3	50.3	18,687	4,674	10.9
9	Male	8,144	5,551	71.0	24.0	68.2	25,509	6,676	8.9
10	Male	14,071	7,960	58.9	12.0	113.1	46,722	8,074	8.7

Note: Rate of return: 5% to 2010, 4% 2011–30, 3% thereafter
Contribution rate (individual accounts), 8%
Average social wage in 1994, 5,000 yuan
Accrual rate, 1%
Defined benefit rate, 24%
Source: World Bank 1996 (technical annex 3).

new system is one of the political economy issues of the change to the proposed pension system.

Financial viability of pillar 1

The financial viability of the pension system is not an issue for funded individual accounts (pillar 2) since those payments will be linked to contributions on an actuarial basis. For pillar 1, however, the state assumes responsibility for providing defined benefits, so the system must be designed to ensure the financial viability of the system in the face of contingencies that are unavoidable over the long term.

There are many options for the design and implementation of pillar 1. Financial projections for pillar 1 under various assumptions about contribution rates and contingencies relating to macroeconomic and demographic developments are shown in table 3.5 and figure 3.3. In the base case scenario, the sustainable contribution rate is 8.3 percent and pension payments from the reformed pillar 1 go only to new retirees (payments to old retirees are covered by the transition account). Receipts exceed payments in the initial years, and the surpluses are invested by public pension fund companies with a minimum rate of return guaranteed by the state. In the base case scenario, receipts begin to fall short of payments by 2032. These deficits can be met from the accumulated reserves that reach 4.1 trillion yuan (at 1994 prices) by 2050. By 2050 deficits begin to decline, and the accumulated reserves are adequate to cover the deficits for at least sixteen years.

However, at an 8.3 percent contribution rate the system is vulnerable to several contingencies. Under the reform scenario society provides only about 40 percent of the target replacement rate, so it is assuming a smaller share of the risk than under the present system. But while compliance and rate of return risks on pension funds are reduced with reforms, the risks are not eliminated. Under the downside scenario in which the compliance rate drops from 85 percent to 75 percent, the participation rate declines from 80 percent to 70 percent, and the rate of returns drops by 1 percentage point from the base case, the sustainable contribution rate increases to 10.4 percent. Thus, to provide for contingencies, a 9 percent contribution rate is recommended for pillar 1. Receipts and payments with baseline

TABLE 3.5

Simulations under various assumptions for pillar 1 of a multipillar system

(billions of 1994 yuan unless otherwise specified)

Outcomes and characteristics	Baseline scenario	Downside scenario	Recommended scenario[a]
System outcomes			
Sustainable contribution rate (percent)[b]	8.3	10.4	9.0
System dependency ratio (percent)			
1995	21	21	21
2000	18	18	18
2010	15	15	15
2020	18	19	18
2030	29	31	29
2040	50	53	50
2050	47	50	47
Receipts[c]			
1995	47	52	51
2000	74	87	81
2010	194	244	212
2020	317	427	346
2030	504	728	549
2040	602	924	655
2050	778	1,287	848
Payments			
1995	5	5	5
2000	19	20	19
2010	70	78	70
2020	181	216	181
2030	490	639	490
2040	983	1,355	983
2050	1,187	1,693	1,187
Balance[c]			
1995	42	47	46
2000	55	67	62
2010	124	166	142
2020	137	211	165
2030	14	89	59
2040	−381	−431	−328
2050	−408	−407	−339
Accumulated balance[c]			
1995	42	47	46
2000	316	365	352
2010	1,567	1,852	1,766
2020	3,909	4,677	4,477
2030	6,823	8,171	8,102
2040	6,817	7,746	9,116
2050	4,081	4,067	7,869
System characteristics (changes effective 1995)			
Real wage indexation	none	none	none
Coverage rate by 2010 (percent)			
Township enterprise	50	50	50
Other enterprises	50	50	50
Benefits (percentage replacement rate)			
Retirement	24	24	24
Disability	24	24	24
Survivor (lump sum)	100	100	100

TABLE 3.5 (continued)
Simulations under various assumptions for pillar 1 of a multipillar system

Outcomes and characteristics	Baseline scenario	Downside scenario	Recommended scenario
Retirement age (by 2040)			
Men	65	65	65
Women	65	65	65
Participation rate (percent)	80	70	80
Compliance rate (percent)	85	75	85
Wage increase rate (including seniority increase)	5% to 2010, 4% 2011–30, 3% 2031 and thereafter	6% to 2010, 5% 2011–30, 4% after 2031	5% to 2010, 4% 2011–30, 3% after 2031
Interest rate	5% to 2010, 4% 2011–30, 3% 2031 and thereafter	4% to 2010, 3% 2011–30, 2% after 2031	5% to 2010, 4% 2011–30, 3% after 2031

a. Baseline with contingencies covered.
b. Rate that ensures that terminal accumulated reserves in 2050 are ten times the annual deficit in that year.
c. Receipts and balances assume a sustainable contribution rate.
Source: World Bank 1996 (technical annex 4).

assumptions and a contribution rate of 9 percent are shown in table 3.6.

The total recommended contribution rate of 17 percent for mandatory pillars for China (13 percentage points from the employer) is in line with rates in East Asian countries such as Japan (16.5 percent, with 8.25 percentage points from the employer), Malaysia (22 percent and 12 percentage points), Singapore (40 percent and 20 percentage points), and Vietnam (15 percent and 10 percentage points). It is also comparable to the simple average contribution rate in OECD countries of 20 percent, with 12 percentage points coming from the employer (see table A3.1 in the statistical appendix). The rate will, however, be significantly higher than in countries such as the Philippines (8 percent), Indonesia (6 percent), and Thailand (6 percent), where pension programs are at an early stage.

There are many other options for pillar 1. The government might consider a lower contribution rate (6–7 percent) over the next five to ten years to allow for near-term system balance and then increase the contribution rate to the long-term sustainable level thereafter. The rationale for this approach is to take into account two possible risks associated with the proposed scheme: First, if contribution rates are high, evasion might become common in the state sector, and noncompliance in the nonstate sector might prevent coverage from

being extended as planned (see table 3.7 for the effects of low coverage rate). Second, because interest rates have not been rationalized and the regulatory framework is incomplete, accumulating surpluses in an untested regulatory and financial environment might result in inefficient allocation and unwise investment of pension reserves. A phased approach to the contribution rate might allow time for nonstate and township enterprises to adjust, for the regulatory framework to be completed, and for administrative institutions to mature.

Similarly, the rate at which the retirement age is raised will influence the contribution rate required (see table 3.7) and the subsequent cash flows for pillar 1. If the coverage rate can be extended only to 20 percent by 2010 and the retirement age does not reach 65 until 2040, the required contribution rate will be 9 percent; it could be as low as 7.3 percent if the retirement age is set to 65 by 2010 and coverage is 80 percent.

Risks associated with a low real interest rate could have an even greater impact on the required contribution rates. If inflation is high and interest rates are not rationalized so that real interest rates are negative, the sustainable contribution rate would have to be as high as 12 percent in order to keep the replacement rate at 24 percent for pillar 1. Alternatively, the government would have to reduce the replacement rate to maintain the financial viability of the public pillar, which is politically difficult. If, however, the real interest rate were 1 percentage point higher than the growth in real wages, the government could lower the contribution rate to 7.6 percent (given a retirement age of 65 by 2040). In the best case, the contribution rate could be set at 6.7 percent, assuming that the retirement age is raised more aggressively and interest rates are rationalized more rapidly (table 3.8).

Options for funding the transition

Annual pension payments owed to current retirees and to current workers for their past service are between 83 and 122 billion yuan until 2039, when they will begin to decline. Still, the period of transition payments is long. Even by 2050, 22 billion yuan of annual payments are due. The present value of these payments is about 1.9 trillion yuan for the whole system and about 0.9 trillion yuan for the state enterprise sector.

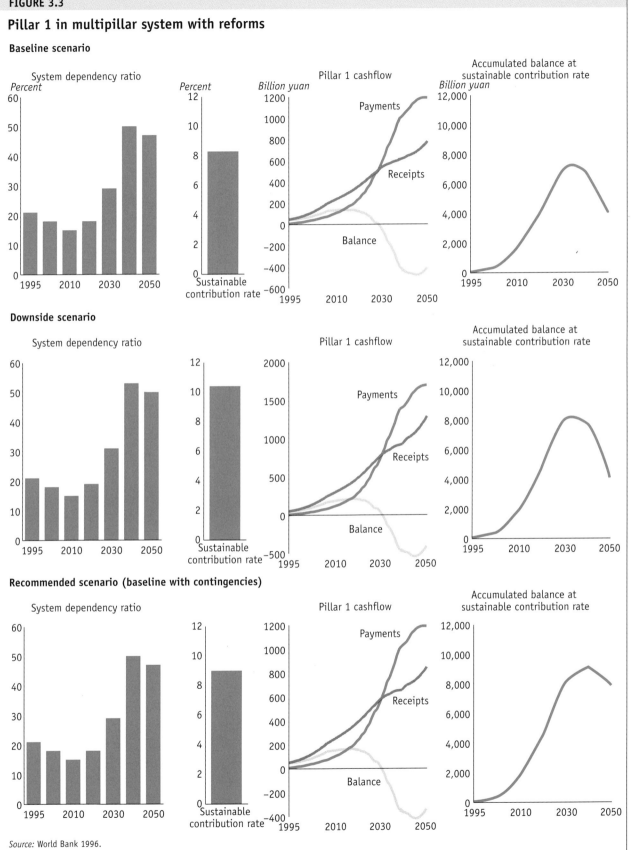

FIGURE 3.3

Pillar 1 in multipillar system with reforms

Baseline scenario

System dependency ratio

Pillar 1 cashflow

Accumulated balance at sustainable contribution rate

Downside scenario

Recommended scenario (baseline with contingencies)

Source: World Bank 1996.

TABLE 3.6

Simulations for pillar 1 in the recommended scenario with a 9 percent contribution rate
(billions of 1994 yuan)

Year	Receipts	Payments	Balance	Cumulative balance with interest	Interest (percent)
1994				0.0	
1995	50.7	4.8	45.9	45.9	5
1996	55.4	7.6	47.7	95.9	5
1997	60.6	10.8	49.8	150.6	5
1998	66.6	14.3	52.3	210.4	5
1999	73.2	17.8	55.5	276.4	5
2000	80.7	19.2	61.5	351.7	5
2001	89.6	23.4	66.2	435.5	5
2002	98.6	27.8	70.8	528.0	5
2003	108.7	32.6	76.1	630.5	5
2004	120.1	37.6	82.4	744.5	5
2005	132.9	40.5	92.4	874.1	5
2006	146.9	46.6	100.3	1,018.1	5
2007	161.4	53.1	108.3	1,177.3	5
2008	176.9	60.2	116.7	1,353.0	5
2009	193.5	66.6	126.8	1,547.4	5
2010	211.6	70.0	141.5	1,766.3	5
2011	224.3	79.1	145.3	1,982.2	4
2012	236.6	88.5	148.1	2,209.6	4
2013	249.0	98.6	150.4	2,448.4	4
2014	261.7	108.0	153.6	2,700.0	4
2015	274.7	116.3	158.4	2,966.4	4
2016	288.5	129.1	159.4	3,244.4	4
2017	302.0	142.8	159.2	3,533.5	4
2018	316.0	155.6	160.4	3,835.2	4
2019	330.8	173.6	157.3	4,145.8	4
2020	345.6	180.7	164.9	4,476.5	4
2021	364.1	201.7	162.4	4,818.0	4
2022	382.1	224.6	157.4	5,168.1	4
2023	401.1	248.2	152.9	5,527.8	4
2024	421.3	279.4	142.0	5,890.8	4
2025	441.4	301.3	140.2	6,266.6	4
2026	464.6	339.8	124.9	6,642.1	4
2027	485.8	375.1	110.8	7,018.6	4
2028	508.7	423.4	85.2	7,384.6	4
2029	529.1	475.1	54.0	7,733.9	4
2030	549.1	490.3	58.8	8,102.1	4
2031	573.9	545.9	28.0	8,373.1	3
2032	588.5	593.4	−4.9	8,619.4	3
2033	603.9	654.9	−50.9	8,827.1	3
2034	614.4	720.1	−105.7	8,986.2	3
2035	622.2	761.7	−139.5	9,116.3	3
2036	635.2	812.4	−177.2	9,212.6	3
2037	645.5	877.5	−232.0	9,257.0	3
2038	649.6	942.4	−292.8	9,241.9	3
2039	652.1	1,002.4	−350.3	9,168.9	3
2040	655.0	983.1	−328.1	9,115.9	3
2041	682.3	1,006.3	−324.0	9,065.3	3
2042	699.1	1,056.1	−357.0	8,980.3	3
2043	707.1	1,095.1	−387.9	8,861.8	3
2044	718.2	1,127.5	−409.3	8,718.4	3
2045	731.9	1,148.8	−416.9	8,563.0	3
2046	749.0	1,168.1	−419.0	8,400.9	3
2047	769.4	1,180.4	−411.0	8,241.9	3
2048	792.9	1,186.9	−394.0	8,095.2	3
2049	819.1	1,188.2	−369.1	7,969.0	3
2050	847.5	1,186.6	−339.0	7,869.0	3

Source: World Bank 1996 (technical annex 4).

TABLE 3.7

Sensitivity of contribution rates for pillar 1 to retirement age and coverage rate: Sustainable contribution rates
(percent)

Coverage rate[a] for township and other enterprises	Retirement age of 65 for men and women by the year			
	2010	2020	2030	2040
20	7.60	7.89	8.25	8.73
30	7.49	7.72	8.05	8.52
40	7.41	7.61	7.90	8.37
50	7.35	7.53	7.80	8.27
60	7.32	7.47	7.73	8.19
70	7.28	7.42	7.66	8.12
80	7.26	7.39	7.62	8.07

a. Assumed to reach the specified level by 2010 and to remain there.
Note: All assumptions are consistent with the recommended scenario except for retirement age and coverage rate. Results would be similar if compliance rates rather than coverage rates declined.
Source: World Bank 1996 (technical annex 4).

The three options

One source of funds for the transition could be increased contribution rates in pillar 1. This would require an additional contribution rate of 3 percentage points. This option would be administratively simple but would have two major drawbacks. It would raise mandatory contribution rates (pillars 1 and 2 combined) to 20.1 percent, which is high by international standards, and would create disincentives for foreign investors and the growth of the nonstate sector in general. Second, it will impede labor mobility, because the host province would bear the pension burden of an incoming worker.

A second option would be to finance the transition by imposing an additional contribution rate on the state sector alone, since it will be the main beneficiary of the pension reform. The additional contribution rate required would be 8.8 percentage points, for a total mandatory contribution rate of 25.8 percent. That rate might be too heavy a burden for the state sector, which is already under financial strain, and the problem of labor mobility remains.

A third approach (as discussed in chapter 2) is an asset-liability swap for enterprises. In this approach, the transition plan would be fully funded from the beginning. Marketable assets equivalent to the implicit pension debt owed by the participating work units would be set aside at the time of reform to fund pension obligations. In market economies, the value of land and

TABLE 3.8

Sensitivity of contribution rates for pillar 1 to real interest rates and retirement age: Sustainable contribution rates
(percent)

Interest rates			Retirement age of 65 for men and women by the year			
1995	2011	2031	2010	2020	2030	2040
−2,	−3,	−3	10.80	11.00	11.20	11.50
0,	−1,	−2	10.30	10.40	10.70	11.00
2,	1,	0	9.30	9.40	9.70	10.10
4,	3,	2	8.00	8.20	8.50	8.90
5,	4,	3	7.35	7.53	7.80	8.27
6,	5,	4	6.70	6.90	7.10	7.60

Note: Wage growth rates are assumed to be 5 percent in 1995, 4 percent in 1995–2010, and 3 percent in 2031 and after. Other assumptions are consistent with the recommended scenario.
Source: World Bank 1996 (technical annex 4).

housing are each approximately equal to GDP. In China, the value of real estate relative to income seems to be even higher. A conservative estimate would put the value of housing and land use rights of the state sector at least equal to GDP. The nonproductive assets of state enterprises (housing and land use rights) are probably large enough to meet their pension debts. With capital gains of 5 percent a year, the annual gain from these assets could be conservatively estimated at 200 billion yuan—more than enough to settle the pension debt. These capital gains do not have to realized in the near term.

The immediate cash flow needs of the transition account can be met by borrowing from pillars 1 and 2, which would have substantial surpluses from the very beginning of reform (see table 3.9). Any borrowing by transition agencies should be explicit, using fully collateralized bonds. Even after meeting the cash flow needs of the transition account, pillars 1 and 2 would have surpluses of more than 150 billion yuan by 2005. These surpluses would be large enough to enable the pension system to assist in the term transformation of savings and contribute to the financing of long-gestation investments in infrastructure and other sectors. After ten to fifteen years, the transition agency will redeem the bonds and pay off the debt owed to pillars 1 and 2 by using either fiscal revenues or the proceeds from selling state enterprise assets. As the ultimate guarantor of pension rights, the government will have to assume the implicit risks on these bonds. However, the contingency provisions in pillar 1 would provide for some cushioning.

The main problems that may arise with the proposed scheme are on the administrative and political side, not on the financial side. Asset-liability offsets always involve complex negotiations about the values of assets and liabilities (in this case, assets in the form of housing and land use rights and liabilities in the form of implicit pension debt), and there would be fears of asset-stripping. However, in the proposed scheme, the initial asset-liability offsets are between state units, and the risks of asset-stripping may not be very high.

Each of the three options noted above has strengths and weaknesses. It is up to policymakers to decide which of these or which combination of options is appropriate for China's conditions. However, it is worth emphasizing that favorable factors today mean that transition costs are more manageable in China than in many other economies that have made similar transitions.

Burden sharing

In addition to illustrative calculations regarding the implications of various methods of financing the transition, a burden-sharing formula is proposed here. Under this formula, current pensioners will be paid through an additional contribution rate of 1.1 percent levied on all covered workers (see table 3.9). To settle the pension debt of existing workers, interest-earning and fully collateralized bonds worth 1,236 billion yuan will be given to the transition agency. Up to the year 2010, the transition agency will borrow from pillars 1 and 2 to cover payments to pensioners over and above the receipts coming from the additional 1.1 percent contribution rate. After 2010 the transition agency will start cashing 10 percent of the bonds. These cash flows will be enough to make transition payments and to start repaying pillars 1 and 2 (see table 3.9). By 2050 receipts from the additional contribution rate will substantially exceed transition payments, and the loans to pillars 1 and 2 will be paid off soon thereafter.

Under this burden-sharing arrangement, the total mandatory contribution rate will become 18 percent. Transition bonds (for all covered workers including civil service and public institution workers) will be 1,236 billion yuan and could safely be collateralized by the housing assets (and the associated land use rights) of the state sector.

TABLE 3.9

Funding for transition account

(billions of 1994 yuan)

Year	Pension to prereform pensioners	Pension for service prior to 1995	Total transition payments	Additional contribution rate (1.1%) for payment to prereform pensioners	Annual borrowing	Accumulated value of borrowings	Surplus in pillars 1 and 2	Annual repayment	Value of transition bonds	Interest rate (percent)
1995	81.4	3.8	85.2	6.2	79.0	79.0	95.4	—	1,236.0	5
1996	77.0	8.0	85.0	6.8	78.2	161.2	103.9	—	1,297.8	5
1997	72.8	12.3	85.1	7.4	77.8	247.0	113.2	—	1,362.7	5
1998	68.7	17.0	85.7	8.1	77.6	336.9	123.6	—	1,430.8	5
1999	64.6	21.3	85.9	8.9	77.0	430.8	135.2	—	1,502.4	5
2000	60.5	22.8	83.4	9.8	73.5	525.9	148.8	—	1,577.5	5
2001	56.6	27.5	84.1	10.9	73.2	625.4	164.0	—	1,656.4	5
2002	52.7	32.3	85.0	12.0	73.0	729.6	179.2	—	1,739.2	5
2003	48.9	37.2	86.1	13.3	72.8	839.0	196.2	—	1,826.1	5
2004	45.1	42.0	87.1	14.6	72.4	953.3	215.3	—	1,917.4	5
2005	41.3	44.5	85.8	16.2	69.6	1,070.6	237.8	—	2,013.3	5
2006	37.8	49.5	87.3	17.9	69.4	1,193.5	260.8	—	2,114.0	5
2007	34.3	54.6	89.0	19.7	69.3	1,322.4	284.4	—	2,219.7	5
2008	31.0	59.8	90.7	21.6	69.2	1,457.7	309.3	—	2,330.7	5
2009	27.7	63.5	91.2	23.6	67.6	1,598.3	336.2	—	2,447.2	5
2010	24.5	64.6	89.1	25.8	63.3	1,741.5	367.1	—	2,569.6	5
2011	21.6	69.8	91.4	27.4	—	1,608.0	384.4	192.9	2,405.1	4
2012	18.9	75.0	93.9	28.8	—	1,487.2	400.4	175.5	2,251.2	4
2013	16.3	79.8	96.2	30.4	—	1,378.3	415.9	159.3	2,107.1	4
2014	14.0	83.4	97.3	31.9	—	1,279.8	431.6	145.3	1,972.2	4
2015	11.8	86.5	98.3	33.5	—	1,190.6	448.5	132.5	1,846.0	4
2016	9.9	91.2	101.1	35.2	—	1,112.2	463.1	118.7	1,727.9	4
2017	8.2	95.6	103.8	36.8	—	1,043.9	476.3	105.9	1,617.3	4
2018	6.7	98.3	105.0	38.5	—	983.9	490.3	95.3	1,513.8	4
2019	5.3	102.4	107.7	40.3	—	933.1	501.7	84.0	1,416.9	4
2020	4.0	101.7	105.7	42.1	—	886.7	536.8	78.1	1,326.2	4
2021	3.0	105.5	108.5	44.4	—	848.3	534.7	68.6	1,241.3	4
2022	2.1	108.9	111.0	46.6	—	817.6	546.2	59.7	1,161.9	4
2023	1.4	110.9	112.3	48.9	—	792.9	558.5	52.8	1,087.5	4
2024	0.9	114.2	115.1	51.4	—	775.2	565.7	45.0	1,017.9	4
2025	0.6	115.3	115.8	53.8	—	762.3	581.9	39.8	952.8	4
2026	0.3	118.3	118.6	56.7	—	755.7	586.5	33.4	891.8	4
2027	0.2	119.0	119.2	59.2	—	753.1	591.3	29.2	834.7	4
2028	0.0	120.8	120.9	62.0	—	755.3	585.9	24.6	781.3	4
2029	—	121.6	121.6	64.5	—	761.4	574.0	21.0	731.3	4
2030	—	118.0	118.0	67.0	—	766.9	597.7	22.1	684.5	4
2031	—	117.7	117.8	70.0	—	767.1	582.9	20.7	634.5	3
2032	—	115.5	115.5	71.8	—	768.5	560.7	19.7	588.2	3
2033	—	113.1	113.1	73.6	—	770.4	522.9	19.4	545.3	3
2034	—	109.8	109.8	74.9	—	772.2	471.7	19.7	505.5	3
2035	—	105.4	105.4	75.9	—	772.8	442.4	21.0	468.6	3
2036	—	100.3	100.3	77.5	—	770.6	405.0	24.0	434.4	3
2037	—	94.6	94.6	78.7	—	764.9	344.5	27.5	402.7	3
2038	—	88.5	88.5	79.2	—	755.7	272.7	30.9	373.3	3
2039	—	82.4	82.4	79.5	—	742.8	201.7	34.4	346.0	3
2040	—	76.3	76.3	79.9	—	725.9	230.4	38.2	320.8	3
2041	—	70.4	70.4	83.2	—	701.8	229.3	44.9	297.3	3
2042	—	64.6	64.6	85.2	—	671.6	176.6	50.4	275.6	3
2043	—	58.7	58.7	86.2	—	635.9	127.1	55.0	255.5	3
2044	—	53.0	53.0	87.6	—	594.0	89.4	60.1	236.9	3
2045	—	47.3	47.3	89.2	—	545.5	74.9	65.6	219.6	3
2046	—	41.9	41.9	91.3	—	489.8	61.6	71.4	203.5	3
2047	—	36.5	36.5	93.8	—	426.2	63.1	77.7	188.7	3
2048	—	31.3	31.3	96.7	—	354.2	77.3	84.2	174.9	3
2049	—	26.4	26.4	99.9	—	273.4	102.5	91.0	162.1	3
2050	—	21.8	21.8	103.3	—	183.3	134.6	97.8	150.3	3
Present value in 1995	681.3	1,236.3	1,917.6	681.3						

Source: World Bank 1996 (technical annex 4).

Conclusion

The scenarios for the new pension system discussed here will be feasible only if, among other things, coverage is extended to include the nonstate sector and the real rate of return on pension funds is at least equal to the rate of increase in real wages over the long term. That requires unifying the pension system, making it mandatory, and backing it by law and administrative power. It also means developing a system of managing pension funds that gives decent real rates of return on the accumulated surplus—the subject of the next chapter.

Notes

1. For details on the simulation model and results, see World Bank 1996.

2. The sustainable contribution rate is defined as the rate that will meet the financing needs of the pension system until 2050, at which point the accumulated reserves would be enough to cover sixteen years of deficit in the system (the present life expectancy at retirement). The model uses a commutation factor of 10, and the required reserves in 2050 are ten times the deficit. The sustainable contribution rate changes only marginally if this factor is changed. (See glossary for definition of commutation factor.)

3. In the model used for the report, separate pension system calculations are made for government organizations, public institutions, state enterprises, collectively owned enterprises, and township enterprises. However, the basic results are presented for all these groups together. The implicit suggestion is that for the sake of encouraging labor mobility between the government sector and enterprises and reducing the pension burden on the government budget, the proposed pension reform should include the government sector. To the extent that it is considered necessary to give higher replacement rates to government employees (in view of their relatively lower wages), that could be done through employer-sponsored supplementary pensions under the third pillar. Including the government sector in the pension system reform immediately would save the budget a considerable amount of money over the long term compared with the rise in the budgetary burden under the present system.

4. The accrual rate could be somewhat higher than 1 percent to begin with to achieve a gradual reduction in replacement rate. A gradualist scenario was considered with an accrual rate of 1.2 percent reduced to 1 percent over twenty years, the increase in retirement age to be postponed to 2010, and the coverage rate of township and village enterprises initially 10 percent (instead of 20 percent) and rising to 50 percent by 2020 (instead of 2010). The results indicate that the long-term financial viability of the system is not affected. However, the transition debt increases by 120 billion yuan, and the surpluses in the first and second pillars after funding transition payments are lower by about 16 billion yuan per year in the first ten years. The ability of the social insurance system to fund infrastructure and other investments is thus reduced. The choice has to be made by political leaders in light of the balance between the higher economic costs of more gradual transition and its greater political and social acceptability.

5. All yuan figures in this chapter are at 1994 prices, unless otherwise specified.

chapter four

Managing and Funding the Three Pillars

he success of the proposed pension reform depends on accompanying reforms in legal, administrative, and financial systems. Extending coverage and increasing compliance are vital for making the proposed pension system viable in the medium term. Thus it is essential to build a strong legal framework and appropriate administrative institutions. Funded individual accounts will yield acceptable wage replacement rates only if the long-term rate of return on pension funds is at least equal to the growth rate of wages. Thus major improvements are required in China's financial sector. Pension reform will fail if decentralized public and private pension institutions are prevented from competing on an equal footing or if government regulations fail to keep pace with the development of financial market and non-bank financial institutions such as pension funds and insurance companies.

Legal framework and governance structure

The central government should regain the authority in pension provision that it relinquished in the wake of the Cultural Revolution. A new social insurance law provides the opportunity. Drafting of a social insurance law ran into difficulties because many controversial issues had not been resolved, including the reconciliation of plans I and II. The process should be reactivated and accelerated.

A new social insurance law should standardize the basic framework of pension provision, including coverage, retirement age, benefit formula, vesting period, and contribution formula for the mandatory component. It should stipulate a central administrative agency or ministry in charge of enforcement of the law and define the responsibilities of government at each level. And it should provide a legal and regulatory framework for the pension management institutions that will manage the basic pension pillar and the individual accounts. It should clearly specify separate management of the two pillars, including organizational structure, ownership, licensing, capital adequacy, and reserve requirements. Information disclosure and investment limits and other prudential and antifraud regulations should also be established.

Social insurance agency

A social insurance agency or ministry should be established as the central administrative agency of the new pension system, with collateral responsibility for the other social obligations of enterprises (housing, health, unemployment insurance).

This agency would be charged with:
- Enforcing the social insurance law.
- Unifying policymaking on pension provision and administration.
- Licensing pension management institutions and providing continuous monitoring and supervision, together with the Ministry of Finance and the central bank.
- Setting up nationwide programs for institution building and personnel training, and computer networks for pension management.

Social security administrations around the world perform at least five basic functions: enumeration, collection of contributions, maintenance of data, determination of eligibility, and post-eligibility maintenance. A national system of individual identification numbers is important for contribution collection and maintenance and should be applied to pensions and health insurance, as well as taxes. An enterprise tax identification number system is also being established. The State Tax Bureau should assist the social security fund administration in collecting mandatory pension contributions. Banks should be required to record individual and enterprise tax identification numbers for opening accounts, and pension administrators should collaborate with banks and tax bureaus in monitoring compliance and detecting and punishing evasion.

Social security fund administration

Pension funds should be administered by a separate agency, which should report to the social security agency or ministry or to the State Council. Funds for the basic pension (pillar 1) and for the fully funded individual accounts (pillar 2) should be administered separately. Counterpart agencies should be set up in the provinces and municipalities. Individual accounts should be managed by licensed pension fund management companies in a decentralized fashion (see below).

China's state-owned commercial banks play an important role in record keeping and payments of the pension system. They usually collect enterprise contributions directly through enterprises' bank accounts, on the basis of their wage bill. They are also becoming the channel for payments of benefits to retirees. This system should be continued and strengthened by assigning custodian banks to pension funds. Enterprises should transfer pension contributions (for all the pillars) to the account of the social security agency in the banks, notifying the State Tax Bureau, social security fund administration, and individuals. The State Tax Bureau should enforce compliance. For pillar 2, individual workers should notify their bank and the social security fund administration of their choice of pension fund company. The bank would then transfer funds from the individual account to the designated pension fund company, with notification to the individual worker. The pension fund company would then enter the transaction into the worker's individual account passbook.

Administration of pillar 1 and the transition plan

According to the projections presented in chapter 3, the pension system will have a positive cash flow for the next thirty to thirty-five years, although individual pillars will have different degrees of surplus and deficit over the years. Pillar 1 will have a surplus until 2030 and then begin to run a deficit. If the surpluses of earlier years are invested properly, pillar 1 will have enough reserves to meet the projected deficits. The transition plan will have large deficits until funds are injected through borrowing or sales of state-owned enterprise assets. Pillar 2 would have surplus cash flows until 2030, after which it will draw down its reserves.

Funding the transition program

The main funding problem for pillar 1 in the near term is achieving balance among different pools across the nation. For the transition pool the problem is to achieve balance not only across regions but also across time, particularly where there are large deficits in the initial years. The proposal here is for the transition fund to borrow from pillars 1 and 2 and to use bonds to meet its cash flow needs. The transition pool would pay the same real interest rate as that paid on long-term government bonds. Any remaining funding for this transition will have to come from fiscal sources or sales of the assets of state enterprises.

Level of pooling

One key decision on pillar 1 is the level of pooling of resources. China has been moving during the past ten years toward municipality-based pooling. This has not worked effectively, in large part because municipal pools are too small to spread risks adequately. Many municipalities are saddled with a large number of old enterprises and are unable to meet their pension obligations without increasing contribution rates to unsustainably high levels. The government has proposed broadening the unit of pooling to the prefecture level (between provincial and county governments). But prefecture-level pooling may still not be large enough for effective risk pooling, and prefectural governments often lack the structure and human resources needed to administer pension pooling.

Pooling for pillar 1 ought to start at the provincial level in order to provide pools large enough to spread the risk. Province pooling will also facilitate the incorporation of township enterprises into the system. Since pillar 1 pensions will be calculated as a percentage of average provincial wages, funds will need to be adjusted at that level.

Provincial pooling does not mean that the funds collected under pillar 1 have to go to the provincial authorities. The basic unit of administration could remain the municipality. Municipal governments will be particularly important if sales of state enterprise assets are a large part of transition funding, because the exchange of pension liabilities for some state enterprise assets will be more clearly understood and acceptable at the municipal level. If a significant part of funding for the transition is provided from national fiscal sources (taxes or borrowing), national level pooling will be needed. Receipts and payments can be tallied at the municipal level, with the provincial authorities dealing only with surpluses and deficits at the municipal level. Transfers would compensate municipalities whose system dependency ratio exceeds the provincial average. A certain percentage of payroll collections could be transferred to provincial authorities to cover administration costs and to make these transfer payments.

With pooling at the provincial level, there will be no need for any regular transfers from national authorities. Natural disasters or financial difficulties of particular enterprises, may however occasionally necessitate temporary or long-term assistance to provincial funds. Such assistance should be provided through a national pension fund financed by appropriating a small percentage of pillar 1 contributions from the localities.

A theoretical case can be made for national pooling for basic pensions (pillar 1) under the formal pension system. National pooling would improve labor mobility and provide a level playing field for enterprises in different localities. The different pension needs of provinces would be taken into account in determining center to local transfers under the intergovernmental revenue sharing arrangements that are being developed in China. The exact pace of development of such a national program will depend on political and administrative considerations and can be decided only by the authorities at various levels of government.

Financial management and regulation of pillar 2

The funded individual accounts (pillar 2) are unlikely to provide a credible guarantee of old age security unless the financial system is well enough developed to ensure real rates of return on pension funds close to the growth rate of real wages. For this reason, many policy analysts, especially those looking at transition economies in Eastern Europe (Holzman 1994), argue that financial sector development must precede the move from a pay as you go to a funded system. Unless concerns about the real value of pensions are satisfied, the switch to a funded system may not be politically acceptable. China has several possible mechanisms for providing positive real interest rates on pension funds, and further development of the financial sector will also help to achieve higher rates of return on pension funds. Thus, pension system reform should go hand in hand with financial sector reform and capital market development. There is a clear synergy between pension funds and capital market development (see box 4.1 on Chile's experience).

With prudent investment policies, pension funds can earn real rates of return of 3–8 percent a year. In Malaysia real rates of return averaged 5.2 percent during 1986–94 (box 4.2 and table A4.3), and elsewhere real rates of return have reached as high as 14 percent during 1982–94 (table A4.2). However, inflation can erode the real value of pension funds and careful monitoring is required.

Adequate real rates of return

Pension funds in China face a challenging task in obtaining real rates of return that match or exceed wage growth, which has been high in recent years (5.4 percent a year during 1980–93). Wage growth is likely to settle down over time, however, and a long-term rate of return on pension funds of about 5 percent a year in real terms should be adequate.

Several indicators suggest that the real rate of return on capital is comfortably above 5 percent in China. An analysis of historical rates of return found average rates of return for 1980–85 for four sectors (industry, construction, transportation, and commerce) to be 14 percent (table A4.4a; Chow 1993). More recently, Yu (1995) estimated pretax profit to asset ratios at 13 percent for 1986–95. World Bank–assisted investment projects in China have a consistent record of earning more than 10 percent real rates of return. On an aggregate basis the incremental income-investment ratio (linked at the macro level to rate of return on capital) was

---

BOX 4.1

How Chile's pension funds stimulated capital markets

Chile's pay as you go pension system was in deep financial trouble by the early 1980s with deficits amounting to 5 percent of GDP. Starting in May 1981 Chile replaced its system with a fully funded scheme based on mandatory individual capitalization accounts that are government regulated but privately managed by specialized Pension Fund Administration companies (AFPs).

The pension fund system has grown steadily and has played an important role in capital market development in Chile.

- It is a major source of private savings, accounting for 18.8 percent of national savings in 1990 and 35 percent in 1994. Contributions are increasing rapidly, and the net increase in pension fund assets exceeded 10 percent of GDP in 1994. Total funds accumulated in individual accounts grew at a real rate of 40 percent a year, reaching a stock of $22.3 billion by July 1994, the equivalent of 43 percent of GDP.
- At the end of 1994, the pension system held 55 percent of state securities (Treasury and central bank bonds), 59 percent of

corporate bonds, 62 percent of mortgage bonds, and 11 percent of corporate equities—a significant presence in financial markets.
- Since 1990 pension funds have been allowed to own foreign assets and have begun to diversify into international capital markets. By the end of 1994 about 3 percent of the total assets had been invested outside of Chile—the amount is now over 30 percent.

Chile's success in pension reform shows what domestic nonbank savings institutions (pension funds and mutual funds) can do in aggregating small private savings, supporting domestic capital markets, and raising aggregate savings. As these savings pools have expanded and come to own a large share of domestic assets, they have been allowed to diversify abroad. This allows hedging for currency risks, enhances diversification, and offsets other capital inflows into Chile. This has represented an important expansion and deepening of Chile's capital markets and further integration into global markets.

Source: Shilling and Wang 1996 (p. 90); Vittas 1995a.

higher in China during 1980–95 than in Chile during 1980–95 and in Malaysia during 1986–95, when these two countries had rates of return above 5 percent on their financial investments (table 4.1).

China's economic fundamentals seem to be strong enough to make a 5 percent real rate of return on pension funds feasible in the long run. To do so, however, will require appropriate financial policies and institution building efforts to achieve a positive real interest rate, diversified portfolio distribution, and appropriate fund management and regulation.

In recent years real interest rates have been negative because of double-digit inflation (see tables A4.5, A4.6, and A4.7). For twenty-eight of the thirty-two years between 1953 and 1983 the real interest rate for one-year time deposits was at least 3 percent; real rates were negative for the other four years (Yu 1995). In response to the recent inflation, the authorities provided three- to five-year bank deposits and for a while selectively issued of treasury bonds at inflation-indexed rates. The system of indexation was nontransparent, however, and could ensure only a zero real rate of interest. Moreover, the

---

BOX 4.2

How pension reforms benefit infrastructure development in Malaysia

The Employees Provident Fund, established in the early 1950s, is Malaysia's most important social security institution. By the end of 1994, the fund had 7.3 million members, about half of them active contributors. Participation rose steadily so that by 1994 47.3 percent of the labor force were contributors. The fund is financed by mandatory contributions from employers and employees. Contribution rates began at 5 percent each between 1952 and June 1975 and increased several times to 10 percent for employees and 12 percent for employers (since January 1993). The system has encouraged savings, with its contribution to gross national savings rising from 11.7 percent in 1988 to 14.6 percent in 1994. Tax arrangements are generous: contributions, interest income and capital gains, and retirement withdrawals are all exempt from income and estate taxes.

The Employees Provident Fund is a statutory body under the Ministry of Finance. Its investments are managed by the Management Investment Committee. In the initial years, investment was confined almost exclusively to Malaysian government securities, and lending was confined to loans guaranteed by the government or by banks. Investment in equities was basically for long-term holding with no trading. Since investment was deregulated in 1991 and further in 1995, the share of investments in government securities has been declining and shares in money markets, bonds, and equities have been increasing dramatically (see chart). Real rates of return averaged 5.2 percent from 1986 to 1994.

In the past, the fund assisted infrastructure development indirectly by investing in government securities. More recently, the fund's board has begun to invest in land, properties, and infrastructure projects throughout the country through both debt and equities and in the process has contributed to the development of domestic capital markets. Some recent projects financed by the fund include Kuala Lumpur International Airport (M$4 billion), light rail transit (M$300 million), and the North-

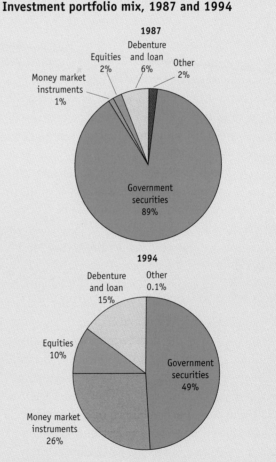

Investment portfolio mix, 1987 and 1994

South Expressway (M$300 million). The government intends to list some infrastructure projects on the local stock exchange, which could boost the liquidity of the fund's equity investments in these projects and provide opportunities for realizing capital gains.

Source: Asher 1995.

TABLE 4.1

Growth of income and implicit rates of return on investment, selected countries
(percent)

Indicator	China	Chile	Malaysia	United States
Annual GDP growth rate (1980–93)	9.6	5.1	6.2	2.7
Investment-income ratio (1980–93)	33.6	20.8	31.8	18.7
Gross incremental income rate of return on investment[a]	28.6	24.6	19.5	14.4
Real rate of return on pension funds	..	14.0 (1981–94)	5.2 (1987–94)	9.5 (1983–93)

a. Defined as: $(\Delta Y/Y)/(PiI/PY) = (\Delta Y/I)*(P/Pi) = (1/ICOR)*(P/Pi)$ where Y = GDP, I = investment, Pi = price of investment goods, and P = GDP deflator.
Source: Staff calculations from data from World Bank 1995f.

indexed bonds were available only to individuals, not to institutional investors. Pension funds had to invest 80 percent of their accumulated surpluses (anything over two months of pension expenditure) in treasury bonds and 20 percent in bank deposits, both at negative real rates of return. In the past few years the 44 billion yuan accumulated in these funds has been depreciating in value.

While workers bear the investment risks under defined contribution schemes, the government can guarantee a minimum real rate of return on pension funds. One way is by issuing treasury bonds indexed to inflation. The government of Chile, for example, provides several kinds of guarantees for its privatized pension system. The pension institutions are responsible for ensuring a minimum real yield for the pension funds they have administered over the previous twelve months (based on the average yield of all pension funds over the same period). Should a pension management company fail to produce the minimum yield, the government, after pursuing all other responses stipulated by the law, will make up the difference and liquidate the company. In addition, the state guarantees certain minimum benefits in the event of the bankruptcy of a pension management company. Since China's proposed multipillar system would have a public pension provision under the first pillar, public guarantee of pension benefits for the second pillar should be avoided (and would give rise to problems of moral hazard and adverse selection). Instead, a minimum real rate of return can be used as a regulatory tool to monitor pension institutions and identify nonperforming pension funds.

Clear investment rules

The investment rules established for pension funds form the link between pension reform and financial market development. These rules need to be adapted to the growth and maturity of capital and real estate markets. Tight investment limits to ensure adequate risk diversification are justified in countries where well-functioning capital markets are absent, prices are volatile, and fund managers lack experience. In Malaysia and Singapore most pension funds are invested in government bonds and other debt instruments, with only a small proportion in equities. But both Malaysia and Singapore allow individual workers to invest their provident fund balance in housing and other approved securities, and as financial systems have matured, workers have been given more control over their investments. Chile initially subjected pension funds to strict rules with upper limits on investment in various instruments and issuers (companies). The limits were gradually relaxed in line with the growing maturity of capital markets (box 4.3). Equity investments, initially forbidden, have been gradually allowed since the mid-1980s (the current limit is 37 percent of total assets).

Diversification of risks

Pension funds must diversify their portfolios to reduce risks. In industrial countries pension fund portfolios are well diversified in a large variety of instruments—short term and long term, equity and debt, public and private, domestic and international, and financial and real estate (tables A4.8–A4.10). China must allow pension funds to do the same.

In the short to medium term (one to five years) before interest rates are rationalized, capital markets are integrated and opened, and regulation of nonbank financial institutions is completed, both upper and lower limits should be set on pension investments (see table 4.2 for specific suggestions). Lower limits should be used only as a transitional measure, to ease concerns about state banks losing funds. Lower limits can create incentives for governments to pay low rates of interest on their bonds and for state banks to ignore efficiency improvements. Lower limits also make it difficult to hold fund managers accountable for profits and losses, since they

How Chile adjusted investment rules as its capital markets deepened

Investment rules must be adjusted to the growing maturity of the financial market and pension system. Initially, pension investments in Chile were limited to public sector securities (treasury and central bank liabilities), bank liabilities, mortgage and corporate bonds, and quotas of other pension funds. The upper limits were 100 percent for state securities, 80 percent for mortgage bonds, 70 percent for bank liabilities, 60 percent for corporate bonds, and 20 percent for the quotas of other pension funds. No lower limit was imposed. The following changes were made in subsequent years:

• 1982: The limit on bank liabilities was reduced to 40 percent.

• 1985: The limit on state securities was lowered to 50 percent and that on corporate bonds to 40 percent. Investments in equities of privatized state enterprises were allowed up to 30 percent. A limit per issuer was imposed equal to 5 percent of the value of the fund or of the capital of the issuer.

• 1986: Authorization to invest in equities was extended to corporations set up as joint stock companies, provided no individual shareholder held more than 20 percent of the equity capital (requiring dispersed ownership).

• 1988: The ownership concentration limit for investment in corporate equities was raised to 50 percent, and the use of concentration factors was introduced.

• 1989: Investment in real estate companies was authorized subject to an aggregate limit of 10 percent of the fund and an individual limit of 7 percent of the fund or 20 percent of the capital of the company. (Real estate companies were allowed to grant mortgage loans and invest in mortgage bonds or mortgage-backed securities.)

• 1990: Pension funds were authorized to invest in commercial paper up to 10 percent of the value of the fund, shares of investment funds up to 10 percent, and foreign securities, with an initial 1 percent limit to be raised by a percentage point over the next three years and to 10 percent in the fifth year.

Source: Vittas and Iglesias 1992.

have little decisionmaking power. The current investment rules set by the Chinese government suffer from both these incentives problems.

In the immediate term the government needs to issue and develop a range of financial instruments, including more long-term (five to ten years) treasury bonds. A secondary market for treasury bills and bonds has been established, but actions are needed to expand capitalization and increase liquidity to meet the needs of pension fund investors. Banks must be allowed to issue long-term certificates of deposit designated for institutional investors in large denominations and with penalties for early withdrawal. Secondary market for certificates of deposit will need to be developed as part of financial market development. Infrastructure construction bonds with a maturity longer than fifteen years should also be issued.

In the medium term the government can gradually allow pension funds to invest in corporate equities and bonds. A low upper limit should gradually increase by 2 percentage points a year to reach 10 percent in five years. Equity shares must be in blue chip companies, and corporate bonds must be rated higher than investment grade. Credit rating agencies also need to be developed. Pension fund participation would provide more liquidity in capital markets and would accelerate the

TABLE 4.2

Illustrative investment rules for pension funds
(percent)

Instruments and assets	Upper limit	Lower limit[a]
Government bonds (3 + years)	80	20
Government bonds (3 years or less)	50	0
Bank certificates of deposit or deposits (3 + years)	50	20
Bank certificates of deposit or deposits (3 years or less)	30	0
Infrastructure bonds (guaranteed)	20	0
Corporate bonds (above investment grade)	2–10	0
Equities in blue-chip corporations	2–10	0
Commercial real estate	10	0
Residential housing (mortgage bonds)	10	0
Foreign government, corporate bonds	0–1[b]	0
Foreign equities	0–1[b]	0
Anticoncentration rule		
Pension fund share in a single company	10	0
Maximum holding of a company's capital	10	0
Regional diversification rule		
Share of fund assets invested within own city or prefecture (excludes bank deposits)	20	0
Share of fund assets invested within own province[c]	40	0

a. Lower limit should be eliminated and upper limits raised once financial markets have deepened and regulations are more complete.
b. This limit reflects the fact that China's currency is not fully convertible and the capital account is closed. It should be raised in the long run.
c. Applies to four megacities, Beijing, Shanghai, Tianjin, and Chongqing.

issuance of initial public offerings and the expansion of market capitalization.

On real assets different policies should apply to commercial real estate and noncommercial residential housing. The principle here is to allow pension funds the freedom of selection and to provide liquidity, transferability, and risk diversification. A market for commercial real estate already exists, and returns are reportedly high. Pension funds should be allowed to invest and reap the high capital gains. They could be allowed to hold either the real assets or the mortgage-backed securities, provided the buildings are already completed and marketable. There is as yet no market for noncommercial residential housing, and the housing stock is not liquid. Pension funds may be allowed to hold mortgage-backed securities issued by housing management companies on a voluntary basis. Workers residing in the building would be ensured user rights for one generation, but the pension fund would hold the collateral right. A market for mortgage bonds should be developed so that the bonds could become liquid and tradable on the secondary market.

International diversification is important for hedging exchange rate and other risks. Until the currency becomes fully convertible and restrictions on capital account transactions are lifted however, the upper limit on foreign assets is set at 0–1 percent.

Rules against concentrations of holdings in a single company or issuer could be expressed in two forms. One could be a percentage of fund assets that could be invested in a single company (for example, 10 percent). If a pension fund is big and the company small, this rule will not be binding in most cases. A second rule could be expressed as a percentage of the company's capital (say, 10 percent again). This rule will prevent pension funds from holding a controlling stake in a company and would put an upper limit on potential losses in the case of bankruptcy. The rule will be binding when the company is small. The rule can also be used to guard against self-investment in management-related private businesses.

China is a country with wide income disparities, and investment opportunities and returns are not located evenly across provinces. Provincial and local governments tend to keep the investable pension funds within their own regions, which will prevent pension funds from spreading risks adequately and reaping higher returns. A regional diversification rule should aim to reduce the fragmentation of capital markets, promote investment flows across regions, and allow capital to flow to where it is needed the most. The limits are set generously at 20–40 percent and apply only to local infrastructure, local corporations, real estate, and residential housing.

Fund management and regulation

The fully funded individual accounts of the second mandatory pillar should be managed separately from the basic pension pillar. Funds accumulated in individual accounts should not be used to finance pension payments for pillar 1 or the transition account. An explicit borrowing mechanism through bond issuance must be established between different pillars to facilitate risk sharing between older and younger generations and to ensure loan repayment. This can be achieved only if the two pillars are managed by different pension institutions.

The first pillar can be managed by social insurance companies directly affiliated with the Bureau of Social Insurance. The second pillar should be decentralized and competitively managed by licensed, corporatized state, joint venture, or private investment companies through competitive bidding. These companies must be licensed by the government as pension fund management companies and have chartered financial analysts or investment officers. Qualification rules for the companies should be set in accordance with the law and regulations for nonbank financial institutions. Certification exams should be organized for chartered financial analysts.

It is essential that the fund management companies be decentrally managed and operate in a competitive environment. Both state and nonstate management companies should be allowed to operate in competition with each other, as is the case in Argentina, where a government-run pension fund is allowed to compete with several private pension funds and workers are able to select among these funds.

International experience suggests two options for selecting fund management companies. Local governments can select the companies by organizing public bidding for the contract and specifying bidding criteria. A locality may select several companies to manage the funds in individual accounts. Separate trust accounts should be established for pension funds in each of the companies, and the accounts should be managed independently of the other business of the company. Another option is for individual workers to select pension fund management companies by taking their individual accounts to a licensed investment company. Workers would notify their employer and the pension fund administration of the company that is handling their individual accounts. They will be allowed to switch companies periodically, taking into account the performance of these companies.

Competition as a regulator. The second option, individual selection, has several advantages. It allows individual workers to monitor the performance of pension funds and to switch companies when they are dissatisfied. Thus competition among management companies can serve as a regulatory and monitoring mechanism. Second, it avoids the difficulty of having to select from among many bidders with no track record of managing pension funds. Third, it may avoid the conflict of interest inherent in awarding and enforcing contracts by a municipal government to companies that are often owned, at least partially, by the government. The administrative costs, however, are likely to be high under a system that allows individual-based selection. An alternative might be to have local governments select management companies and, at the same time, to create a competitive market for fund managers and encourage the hiring of experienced foreign or non-local fund managers—which could reduce the problems of conflict of interest.

Prudential regulations. Pension fund management companies must meet capital adequacy requirements and reserve requirements, which may be set by the central bank. Because companies are fully accountable for the profits and losses of the funds they manage, they must be subject to regulations such as investment and antifraud rules and requirements for information disclosure and portability and must be required to act in the best interests of their beneficiaries or shareholders. Quantitative restrictions on asset allocation, as discussed above, are also important for promoting asset diversification and protecting pension funds. Contracts to manage pension funds could have three- to five- year terms, with incentives designed to lower administrative

costs and increase investment returns. Fund managers and management firms should be evaluated annually, and their compensation should be linked to their performance, compared across time and against the industry average. Incentives should be designed so that fund managers want to acquire reputations for good performance and are penalized for poor performance.

Disclosure of information. Pension funds are owned by contributing workers. These workers have a right to regular reports (semiannually or quarterly) on how pension funds are invested and what their rates of return are. Information disclosure should be covered by the Social Insurance Law. Annual reports should be audited and made available to pension fund members free of charge. Reports should provide information on the names of actuaries and fund managers, number of beneficiaries, amount of contributions, increases in benefits to current pensioners, distribution of assets, actuarial statement on financial viability of the fund, and the performance of fund managers and how they are remunerated. Every three years a more detailed report should examine longer-term financial viability. Information must also be presented in a standardized form that is easy to understand. For instance, investment rates of return for three months, one year, and five years should be presented along with sector averages to allow for comparison of fund managers' performance.

Portability. The Social Insurance Law should include provisions on the vesting and portability of fully funded individual accounts and employer-based supplementary plans (third pillar) that do not impede labor mobility. For example, certain banks could be designated as custodian banks, to take care of the individual accounts of workers who are looking for a new job or moving to another region.

Taxation. In pension systems around the world tax incentives are often used to encourage compliance. Pensions may be taxed when contributions are made, when investment income is earned, or when retirement benefits are paid. In general, there are two types of tax treatment. One exempts contributions and investment income from taxes, but taxes pensions. The other taxes contributions, but exempts investment income and pensions.[1] In China, Document 6 of 1995 stipulates that contributions to the basic social pension are tax exempt, but it is not clear whether the same treatment is given to contributions to the supplementary pension schemes. This is a complex issue with significant fiscal impact and needs to be examined in conjunction with tax and expenditure reforms.

Regulatory structures and procedures. The Social Insurance Law should stipulate the government agencies responsible for supervising and regulating the pension fund management companies. Monitoring and supervision should be close and continuous. Annual reports, accounts prepared by auditors, and actuarial reports should be required, and computer checks and on-the-spot inspections of pension funds can also be used. Balance sheets and other financial statements should be audited and examined periodically by supervisory agencies. Since the regulatory framework is still underdeveloped, the government can delegate some supervision to self-regulatory bodies within the industry, such as professional associations of fund managers.

Development and regulation of pillar 3

China needs to develop its institutional and regulatory capacity for providers of supplementary pensions: insurance companies and employer-sponsored pension funds. Two major impediments have slowed the development of supplementary schemes. The high replacement rate in the current system has reduced incentives on the demand side, while the absence of a legal framework for pension funds has impeded development on the supply side. In China the only companies providing annuities are commercial life insurance companies. In other countries, private employer-sponsored pension funds also provide annuities, through defined benefit or defined contribution plans. By enacting the Social Insurance Law and completing the regulatory framework for pensions funds, China could promote the establishment of employer-sponsored pension funds in a manner that is consistent with international norms.

Overhauling the insurance sector

There was no commercial insurance in China in the two decades before 1979, but the industry has experienced spectacular growth since then. Total premiums rose from 2.6 billion yuan in 1985 to 37.6 billion yuan in 1994, an annual growth rate of 135 percent. Still, total premiums accounted for only 0.84 percent of GNP in 1994, and life insurance for only 31 percent of total premiums (table A4.11). By comparison, insurance premiums were 12 percent of GNP in the Republic of Korea, nearly 4 percent in Malaysia, and 3 percent in Chile in 1992. Thus China's insurance industry is still underdeveloped.

The industry has been dominated by the People's Insurance Company of China (PICC), a state-owned entity tightly controlled by the government (tables A4.11 and A4.12). Domestic competition was recently introduced, and there are now more than thirty authorized insurance companies in China, eleven of which are operating at the national level. PICC still held nearly 70 percent of the insurance market in 1993 measured by premium revenues (31.15 billion yuan). And through its controlling shares in regional insurance companies, PICC effectively controlled as much as 85 percent of the market in 1994. Two foreign insurance companies have been allowed to do business in China, the American International Group (AIG) and Tokyo Marine and Fire Insurance. A Sino-Canadian joint venture insurance company was approved in May 1996.

Remedying the investment structure mismatch. The investment policies of insurance companies, in particular PICC, have been tightly controlled by the government. A mismatch between investment structure and the structure of insurance funds has resulted in long-term funds being invested in short-term assets. Since 1988 PICC has been allowed to purchase financial bonds, to lend to banks, and to make loans to enterprises for working capital and for technology upgrading and innovation. It is prohibited from making loans on fixed capital construction. Reverse term transformation is occurring since life insurance reserves are long term in nature but are being forced into short-term investments. As a result, the asset allocation of Chinese insurers diverges greatly from international standards (see, for example, the asset distribution of U.S. life insurers, figure A4.1).

Irrational investment policies have led to the misallocation of capital, reducing economic efficiency and causing low or even negative rates of return for PICC.

PICC's interest income on life insurance amounted to 1.6 billion yuan in 1993 on an estimated 20.3 billion yuan available for investment throughout the year. Given an inflation rate of 13.2 percent for the year, the real rate of return was –5.3 percent.

Promoting domestic competition. To provide opportunities for the insurance sector to develop and to promote domestic competition, PICC's near monopoly should be ended. China's experience with its airline industry shows how breaking up a monopoly and introducing domestic competition can produce spectacular growth and benefit consumers through better services (box 4.4). For the insurance industry the first step might be to allow certain branches to become independent firms and then to allow all firms to engage in an overlapping range of businesses in all regions. The Insurance Law, issued in October 1995, stipulates the separation of the life and nonlife insurance businesses. PICC has been transformed into a shareholding corporate group, and life insurance, nonlife insurance, and reinsurance are handled by separate subsidiaries. These are steps in the right direction, but they may not be sufficient.

New domestic companies need to be encouraged as well, especially in life insurance. The relationship between banks and nonbank insurance companies will need to be clarified by law, so that the industry can develop in an orderly fashion. Since more financial and investment experts are working in the banking sector, it may be practical to allow banks to participate in the insurance business through bank holding companies. More foreign and joint venture insurance companies should be allowed into the market, both to stimulate competition and to bring in needed actuarial and finan-

---

BOX 4.4

A domestic competition success story: Civil aviation

In October 1987 China Civil Aviation, a giant state-owned airline—China's only airline at the time—was broken into eight state-owned airlines. The breakup was organized first along product lines and allowed for regional specialization. Today, there are thirty-nine airlines (including twenty-seven air-transport companies). All the companies are state-owned corporations, and two are listed on the stock exchanges. Some of the airlines are now allowed to fly the same routes. This competition, although it is between state-owned corporations rather than between private companies, has led to spectacular growth in the industry.

Domestic passenger and freight traffic have been growing at a much faster rate than international and regional routes.

Before reform the annual growth rates were 13.6 percent for both passenger traffic and freight; after the reform (1990–95) annual growth soared to 22 percent and 20 percent. To some extent, this growth surge reflects the influence of competition on domestic routes, although demand growth and airport construction are also important factors. While efficiency data are scarce, the utilization of aircraft gives a clue: domestic passenger traffic almost tripled in 1990–95, while the number of aircraft increased only 71 percent. Competition has made domestic air travel convenient and affordable to the Chinese people, where once it was accessible mainly to foreign visitors.

BOX TABLE 4.4

Effects of introducing domestic competition in the airline industry

Indicator	1985	1990	Annual annual growth rate 1985-90 (percent)	1993	1994	1995	Annual annual growth rate 1990-95 (percent)
Passenger traffic (millions of person-kilometers)							
International routes	3,817	5,169	6.1	8,624	9,681	11,497	16.0
Regional routes	793	2,113	19.6	3,957	3,541	3,761	11.5
Domestic routes	7,060	15,766	16.1	35,179	41,935	52,872	24.2
Total	11,672	23,048	13.6	47,760	55,158	68,130	21.7
Freight traffic (millions of ton-kilometers)							
International routes	220	438	13.8	945	981	1159	19.5
Regional routes	31	63	14.2	110	102	105	10.2
Domestic routes	164	316	13.1	606	775	966	22.3
Total	415	818	13.6	1,661	1,858	2,230	20.1
Number of civil aircraft	404	421	0.8	646	681	720	10.7

Source: World Bank staff calculations based on *China Statistical Yearbook 1995* and 1996 (p. 521).

cial expertise to build the human capital base through learning by doing. China is already moving in this direction. By the end of 1995 some 77 foreign insurance companies had established 119 representative offices. Also, the number of years required for starting business operations (three years currently) is likely to be shortened.

Insurance companies need clear and transparent investment rules (similar to those shown in table 4.2) to provide more opportunities for portfolio diversification, risk-spreading, and capital appreciation. Life insurance reserves are long-term funds and need to be invested in long-term instruments and real assets that can produce fixed income and capital gains. The investment upper bound should be in line with international standards in the life insurance sector (see figure A4.1).

Regulating employer-sponsored pension funds

Employer-sponsored private pension schemes have several advantages. They are voluntary, and so there is minimal incentive to evade. They are flexible, and so can fit different tastes. And their administrative costs tend to be lower than individual insurance or savings plans, so they are more efficient. Their shortcomings include complexity, uneven coverage, and sometimes underfunding (for defined benefit type). In more and more industrial countries companies are shifting from defined benefit private pension funds to defined contribution plans, often because of the problem of underfunding (World Bank 1994a). Defined contribution plans are automatically funded and more easily transferred, but workers bear most of the investment risks.

When considering development of the third pillar and pension institutions, China should direct its regulations to promoting defined contribution pension schemes. For instance, preferential tax treatment, if any, should be given only to employer-sponsored pension plans that conform to the principle of defined contribution and other government regulations.

Employer-sponsored pension funds must be legal entities distinct from the plan sponsor and administered by trustees with worker representation. Many regulations for these funds are similar to those needed for the pension fund management companies discussed earlier—prudential regulation, investment rules, information disclosure, and regulatory structures. Some specific regulations are needed as well. For example, vesting

and portability are important to ensure labor mobility and economic efficiency. Vesting standards vary across countries. In Indonesia the law stipulates that benefits are vested and portable after one year of service. In the United States benefits are fully portable after ten years of service. In Japan, with the assumption of lifetime employment, vesting takes fifteen to twenty years, so early leavers are penalized. For employer-based supplementary pension plans, however, portability should not be overemphasized. Employers might need to discourage turnover and may need to offer additional pension benefits to retain high-skilled workers, which is productivity-enhancing.

Pension funds are vulnerable to fraud if prudential standards are absent and if there is no regulatory body to monitor the conduct of funds on behalf of members. For example, in the famous Robert Maxwell case in the United Kingdom, pension fund assets were lent to private companies owned by Maxwell or invested in their shares. The assets were lost when the private companies went bankrupt. One way to avoid this danger is to establish independent custodians for securities transfers. In the United States a custodial financial institution must be selected to safekeep all stocks and bonds held by pension funds. The custodial institution must be independent of the board of trustees as well as of fund managers.

In China the Shanghai Stock Exchange has established custodial services to safekeep stocks and bonds for customers; so have stock trading centers in some large cities. The requirement that all pension funds and institutions select an independent custodial service needs to be embodied in law. Other antifraud measures might include limits on self-investments, more frequent monitoring, less employer influence on trustees, better independent actuarial information, and more employee representatives on boards of directors. There is no once-and-for-all solution for regulatory program. Continuous monitoring, vigilance, and fine-tuning of procedures are needed.

The pace of change

The proposed system of pension reform in China has much in common with reforms recently introduced in Chile and Argentina. Some basic similarities and differences are presented in table 4.3.

TABLE 4.3
Pension system reform characteristics: China (proposed), Chile, and Argentina

Characteristic	China (proposed)	Chile	Argentina
Unified system			
Mandatory contributions			
For basic benefits	9% (pillar 1)	General revenue	}27.0
For individual accounts	8%	10%	
Does the system allow for additional voluntary pensions?	Yes	Yes	No
Does the system provide minimum pension?	Yes	Yes	Yes
How is the minimum pension financed?	Employer contributions to pillar 2	General revenues	Solidarity plan
Substitution or complementary to old system			
What happens to the old system?	Phased out	Phased out	Merged
Is current labor force allowed to remain in old system?	No	Yes	No
Is entry into new system mandatory for new workers?	Yes	Yes	Yes
May workers switch back to old system after entering new system?	No	No	No
Incentives for entering the new system			
Recognition bonds?	Yes	Yes	Not applicable
Reduction of contribution taxes?	Yes	Yes	Not applicable
Rise in take home pay?	No	Yes	Not applicable
Collective or individual affiliation			
Is affiliation individual?	Yes	Yes	Yes
May members move freely between pension fund management firms?	Yes	Yes	Yes
Incentives			
Pension fund investment returns?	Yes	Yes	Yes
Advertising?	Yes	Yes	Yes
Commissions?	Yes	Yes	Yes
Role of the private and public sectors			
Does the public sector regulate pension fund management firms?	Yes	Yes	Yes
Does the public sector supervise the industry?	Yes	Yes	Yes
Is there to be a new public supervisory institution?	Yes	Yes	Yes
May the public sector participate in the financial management of pension funds?	Yes	No	Limited
Does the public sector collect pension contributions?	Yes	No	Yes
May the private sector participate in management of funds with existing financial firms?	Yes, but with separate accounting	No	No
Are the following roles left in public institutions			
Custody of financial instruments?	Custodian banks	Central Bank	Central Bank
Risk classification commission?	Central Bank	Cental bank and Securities commission	Risk clarification commission
Insurance industry regulation and supervision?	Central bank	Equities and insurance superintendency	Insurance Superintendency
Management of the old system?	Ministry of Social Insurance	Pension Normalization Institution	Social Security Administration
Management of minimum pensions?	MSI	IDEM	IDEM
Are the following roles left to private institutions			
Insurance management?	Insurance companies	Insurance companies	Insurance companies
Financial risk classification?	Credit rating agencies	Risk classification firms	Risk classification societies
Separation of programs			
Does the new system provide benefits for other than retirement, disability, survival, and minimum pension social security benefits?	No	No	No
How are other social security benefits provided?	Separately, but under a single ministry	Separately	Separately
Is pension fund management firms' capital firmly separated from the public pension fund?	Yes	Yes	Yes
Are pension fund management firms responsible for providing a minimum return according to market conditions?	Yes	Yes	Yes
Are pension fund management firms obliged to maintain reserves to respond to members rights on their pension fund?	Yes	Yes	Yes

Source: Botka 1994 for Chile and Argentina.

Pension system reform involves a major redirection of financial flows and asset entitlements and affects the livelihoods of millions of people at a highly vulnerable stage of their life. There will be major gainers and losers. Enterprises and localities with younger populations will lose and those with older populations will gain through the pooling proposed here. The proposed scheme tries to reduce the conflict by emphasizing individual accounts and defined contribution systems, leaving the redistributive element to the basic pension component.

Even more important is the transition plan, which may involve major changes in the entitlement to assets. The treatment of women also changes. Women get access to a basic benefit that is independent of their wage and, to that extent, they gain since women tend to be lower wage-earners. Women lose, however, in the extension of their retirement age to match that of men. The pension reforms thus impinge on many powerful political economy considerations, which may play a decisive role in policymaking, but are beyond the scope of this report.

This report emphasizes the urgency of pension system reform in China, an urgency that derives primarily from the need to delink social welfare responsibilities from state enterprise management so as to accelerate state enterprise reform.

Although it is clear that the costs of transition will rise with time, reforms should not take place in an atmosphere of crisis. The pension system reforms involve many risks. If the financial sector and capital market reforms do not materialize, the system of individual accounts will not function properly. Similarly, if system coverage cannot be extended or compliance rates decline or local authorities do not cooperate fully, the reforms will fail. A careful program of consensus building is needed along with mobilization of funds. The Social Insurance Law is urgently needed to standardize the basic framework of pension provision and to establish the institutional infrastructure of the system and the regulatory framework for fund management companies. Experiments with the new system in selected localities should be started soon, along with careful monitoring of the results. These results should then guide the design and implementation of the unified system of pensions, by 2000 at the latest.

Note

1. In the case of employer-sponsored supplementary pensions, the tax incentives for employers are important, and thus many industrial countries give incentives to funded company pension schemes by providing partial tax exemption.

Glossary

Actuarial fairness. A method of setting insurance premiums according to the true risks involved.

Adverse selection. A problem stemming from an insurer's inability to distinguish between high- and low-risk individuals. The price for insurance then reflects the average risk level, which leads low-risk individuals to opt out and drives the price of insurance still higher, creating strains on insurance markets.

Average effective retirement age. The actual average retirement age, taking into account early retirement and special regimes.

Benefit rate. The ratio of the average pension to the average economywide wage or covered wage.

Commutation factor. The present value of an annuity of a unit payable annually in advance from retirement age, contingent on the survivorship of the recepient. The factor is based on the interest and mortality assumptions used in the model.

Contracting out. The right of employers or employees to use private pension fund managers instead of participating in the publicly managed scheme.

Contribution rate. The proportion of workers' monthly wage paid as pension insurance premium to pension fund.

Cost rate. Pension expenditure as a ratio of covered wage bill. This is equal to the system dependency ratio times the benefit ratio.

Defined benefit. A guarantee by the insurer of pension agency that a benefit based on a prescribed formula will be paid.

Defined contribution. A pension plan in which the periodic contribution is prescribed and the benefit depends on the contribution plus the investment return.

Demographic dependency ratio. Ratio of population in age groups 65 years and older to population in age group 15–64 years old.

Demographic transition. The historical process of changing demographic structure that takes place as fertility and mortality rates decline, resulting in an increasing ratio of older to younger persons.

Full funding. The accumulation of pension reserves that total 100 percent of the present value of all pension liabilities owed to current members.

Implicit public pension debt (net). The value of outstanding pension claims on the public sector minus accumulated pension reserves.

Intergenerational distribution. Income transfers between certain age cohorts of persons.

Intragenerational distribution. Income transfers within a certain age cohort of persons.

Legal retirement age. The normal retirement age written into pension statutes.

Means-tested benefit. A benefit that is paid only if the recipient's income and wealth fall below a certain level.

Minimum pension guarantee. A guarantee provided by the Government to bring pensions to some minimum level, possibly by "topping up" the capital accumulation needed to fund the pensions.

Moral hazard. A situation in which insured people do not protect themselves from risk as much as they would have if they were not insured.

Old-age dependency ratio. The ratio of older persons to working age individuals. The old-age dependency ratio used in the text refers to the number of persons aged 65 and above divided by the number of persons aged 15 to 64 (also called population dependency ratio, different from system dependency ratio, noted below).

Pay-as-you-go. In its strictest sense, a method of financing whereby current outlays on pension benefits are paid out of current revenues from an earmarked tax, often a payroll tax.

Pension coverage rate. In this report, the number of workers actively contributing to a publicly mandated contributory or retirement scheme divided by the estimated labor force.

Pension spending. In this report, pension spending is defined as old-age retirement, survivors', death, and invalidity-disability payment based on past contribution records plus noncontributory, flat universal, programs specifically targeting the old.

Portability. The ability to transfer accrued pension rights between plans.

Provident fund. A fully funded, defined contribution scheme in which funds are managed by the public sector.

Replacement rate. The value of a pension as a proportion of a worker's wage (net of individual worker's contribution to the pension system) during some base period, such as the last year or two before retirement or the entire lifetime average wage.

System dependency ratio. The ratio of pensioners receiving pensions from a certain pension scheme divided by the number of workers contributing to the same scheme in the same period.

System maturation. The process in which young people who are eligible for pensions, in a new system, gradually grow old and retire, thereby raising the system dependency ratio to the demographic dependency ratio. In a fully mature system all old people in the covered group are eligible for full pensions.

Total wage bill. Total remuneration payment by employers to employees in a certain period, including bonuses and in-kind payment.

Universal flat benefit. Pensions paid solely on the basis of age and citizenship, without regard to prior work or contribution records.

Vesting period. The minimum amount of time required to qualify for full ownership of pension benefits.

appendix

Statistical Tables

TABLE A1.1

Government expenditure on state sector retirees, 1987–95
(percent)

Year	Share of GNP				Share of total government expenditure			
	Total	Pension	Health	Other	Total	Pension	Health	Other
1987	1.6	1.1	0.2	0.3	8.0	5.6	1.0	1.5
1988	1.7	1.1	0.3	0.3	9.3	6.1	1.4	1.8
1989	1.8	1.1	0.3	0.4	10.2	6.3	1.7	2.2
1990	2.0	1.2	0.3	0.5	10.9	6.5	1.8	2.5
1991	2.1	1.2	0.4	0.5	11.9	6.9	2.1	2.8
1992	2.1	1.2	0.4	0.5	12.8	7.3	2.2	3.3
1993	2.1	1.2	0.4	0.6	14.0	7.8	2.4	3.8
1994	2.2	1.6	0.4	0.3	17.6	12.7	2.8	2.2
1995	2.2	1.6	0.4	0.2	18.7	13.7	3.0	2.1

Source: World Bank staff calculation based on data from *China Statistical Yearbook 1990, 1994, 1995, and 1996.*

Urban employment, 1980–95

Year	Total	Government organization	Public institution	Enterprises Total	State-owned	Collectively owned	Others	Others
Millions								
1980	105.25	4.90	19.12	80.42	—	—	—	0.81
1981	110.53	5.20	19.89	84.01	—	—	—	1.43
1982	114.28	5.77	20.18	86.86	—	—	—	1.47
1983	117.46	5.91	20.57	88.67	—	—	—	2.31
1984	122.29	6.69	20.66	91.55	—	—	—	3.39
1985	128.08	7.18	21.24	95.16	—	—	—	4.50
1986	132.93	7.70	21.76	98.63	—	—	—	4.84
1987	137.83	8.05	22.34	101.75	—	—	—	5.69
1988	142.67	8.43	22.87	104.78	—	—	—	6.59
1989	143.90	8.85	22.76	105.81	72.28	33.23	0.30	6.48
1990	147.30	9.29	22.86	108.44	73.03	33.78	1.63	6.71
1991	152.68	9.45	23.85	111.78	75.02	34.58	2.18	7.60
1992	156.30	9.96	24.06	113.90	76.43	32.34	5.13	8.38
1993	159.64	9.86	24.27	114.36	76.42	32.59	5.35	11.15
1994	168.16	10.28	24.50	113.71	75.45	30.80	7.46	19.67
1995	173.46	10.08	25.35	113.65	75.44	29.45	8.76	24.38
Annual growth (percent)								
1980–90	3.36	6.40	1.80	3.00	—	—	—	21.14
1990–95	3.27	1.63	2.07	0.94	0.65	−2.74	33.61	25.80

— Not available.
Source: World Bank staff calculations, based on *China Statistical Yearbook* (tables 4-1, 4-2, and 4-5).

Number of pensioners, 1980–95

Year	Total	State sector	Government	Public institution	Enterprises
Millions					
1980	8.16	6.38	—	—	—
1981	9.50	7.40	—	—	—
1982	11.13	8.65	—	—	—
1983	12.92	10.15	—	—	—
1984	14.78	10.62	—	—	—
1985	16.37	11.65	—	—	—
1986	18.05	13.03	—	—	—
1987	19.68	14.24	—	—	—
1988	21.20	15.44	—	—	—
1989	22.01	16.29	—	—	—
1990	23.01	17.24	1.316	2.932	12.992
1991	24.33	18.33	1.467	3.244	13.622
1992	25.98	19.72	1.594	3.472	14.661
1993	27.80	21.43	1.772	3.962	15.693
1994	29.29	22.49	1.912	4.102	16.479
1995	30.94	24.01	2.033	4.433	17.263
Annual growth (percent)					
1980–90	10.37	9.94	—	—	—
1990–95	5.92	6.62	8.70	8.27	5.68

— Not available.
Source: China Yearbook of Labor Statistics 1995 and *1996*; China Social Insurance Department, Ministry of Labor.

TABLE A1.4
Wage expenditures, 1980–95

Year	Total	Government organization	Public institution	Enterprises Total	State-owned	Collectively owned	Others
Billions of yuan							
1980	77.24	3.79	9.34	64.11	—	—	—
1981	82.00	4.09	12.31	65.60	—	—	—
1982	88.20	4.58	13.06	70.56	—	—	—
1983	93.46	5.36	13.33	74.77	—	—	—
1984	113.34	6.26	16.41	90.67	—	—	—
1985	138.30	7.87	19.79	110.64	—	—	—
1986	165.97	10.15	24.70	131.12	—	—	—
1987	188.11	11.60	27.90	148.61	—	—	—
1988	231.62	14.04	34.60	182.98	—	—	—
1989	261.85	16.33	41.10	204.42	150.50	50.57	3.35
1990	295.11	19.29	47.77	228.05	168.41	55.05	4.59
1991	332.39	21.75	52.32	258.32	188.72	62.57	7.03
1992	393.92	27.23	59.75	306.94	220.82	70.66	15.46
1993	491.62	35.77	78.68	377.17	271.87	80.00	25.30
1994	665.64	50.55	120.42	494.64	351.75	97.09	45.80
1995	810.00	55.15	137.37	597.43	421.72	112.09	63.65
Annual growth (percent)							
1980–90	13.40	16.27	16.32	12.69	—	—	—
1990–95	20.19	21.01	21.13	19.26	18.36	14.22	52.59

— Not available.
Source: China Yearbook of Labor Statistics 1995 and 1996; China Social Insurance Department, Ministry of Labor.

TABLE A1. 5
Pension expenditures, 1980–95

Year	Total	State sector	Government organization	Public institution	Enterprise
Billions of yuan					
1980	5.04	4.01	—	—	—
1981	6.23	5.00	—	—	—
1982	7.31	5.89	—	—	—
1983	8.73	7.08	—	—	—
1984	10.61	8.16	—	—	—
1985	14.56	11.24	—	—	—
1986	16.91	13.42	—	—	—
1987	20.43	16.42	—	—	—
1988	27.02	20.90	—	—	—
1989	31.33	24.52	2.08	4.44	18.00
1990	38.86	30.61	2.59	5.44	22.58
1991	45.85	36.53	3.20	6.71	26.62
1992	56.86	45.65	4.29	8.90	32.46
1993	74.73	60.09	5.72	12.38	41.99
1994	102.20	86.15	11.07	18.55	56.53
1995	128.38	107.42	11.43	20.24	70.73
Annual growth (percent)					
1980–90	22.70	22.50	—	—	—
1990–95	23.90	25.11	29.69	26.28	22.84

— Not available.
Source: China Yearbook of Labor Statistics 1995 and 1996; China Social Insurance Department, Ministry of Labor.

Local pension fund—revenues and payments
(million yuan)

Area	Receipts				Payments				Reserves end-1995
	1992	1993	1994	1995	1992	1993	1994	1995	
Provinces									
Heilongjiang	1,480	1,900	2,150	2,995	1,330	1,750	1,960	2,840	1,437
Jilin	769	1,179	1,279	1,901	852	1,120	1,279	1,836	667
Liaoning	2,900	3,700	4,600	5,643	2,600	3,400	4,200	5,022	2,682
Jiangsu	—	—	5,005	6,424	—	—	4,994	5,723	2,771
Zhejiang	1,482	2,032	2,938	3,554	1,352	1,845	2,633	2,911	2,275
Fujian	—	—	923	1,493	—	—	762.67	1,250	658
Henan	1,282	1,457	1,745	2,591	9,113	1,183	1,456	2,132	2,289
Hubei	1,145	1,363	1,541	3,020	895	1,145	1,335	2,571	1,922
Guangdong	2,370	3,020	4,107	5,884	1,710	2,220	3,348	4,310	4,731
Hainan	272	442	603	277	213	371	485	275	218
Sichuan	—	—	2,683	4,267	—	—	2,463	4,002	1,768
Yunnan	642	786	978	1,388	576	677	911	1,218	630
Cities									
Tianjin	—	1,064	1,580	2,080	—	1,167	1,541	2,098	407
Harbin	278	360	429	625	236	319	357	638	286
Changchun	157	268	280	387	142	216	266	329	210
Shenyang	638	851	972	1,198	594	864	988	1,114	528
Taiyuan	2.20	1.74	2.24	—	2.24	1.80	2.20	—	—
Wuhan	372	573	534	1,301	364	611	585	1,230	1,320
Chengdu	359	495	497	719	336	433	475	592	401
Chongqing	525	612	727	1,350	453	581	770	1,189	575
Shenzhen	173	418	566	858	44	109	191	420	1,975

— Not available.

Source: Background papers provided by the localities.

TABLE A1.7

Demographic indicators—selected regions and economies, 1990

(percent)

Region and country	Population age			Ratio of women to men		
	60+	65+	75+	65+/15-64	60+/20-59	age 60+
High-income countries						
Germany	20.3	14.9	7.2	21.7	35.2	1.7
Japan	17.3	11.9	4.7	17.1	30.9	1.3
Switzerland	19.9	14.9	6.8	21.8	34.5	1.4
United States	16.6	12.3	5.0	18.7	30.3	1.4
Simple average	18.6	13.6	5.9	20.4	34.0	1.4
Weighted average	18.2	13.2	5.6	19.6	32.9	—
Latin America and the Caribbean						
Simple average	8.2	5.6	1.9	9.5	18.0	1.2
Weighted average	6.9	4.6	1.5	7.6	14.9	—
Eastern Europe and the former Soviet Union						
Simple average	13.8	9.3	3.3	14.6	26.8	1.6
Weighted average	15.3	10.5	3.7	16.4	29.7	—
Middle East and North Africa						
Simple average	6.2	4.0	1.4	6.9	13.5	1.0
Weighted average	5.7	3.6	1.2	5.7	13.4	—
Sub-Saharan Africa						
Simple average	5.2	3.2	0.9	5.9	12.5	1.2
Weighted average	4.6	2.8	0.8	5.6	11.7	—
Asia						
China	8.9	5.8	1.8	8.7	16.6	1.1
Malaysia	5.7	3.6	1.1	6.3	12.5	1.1
Singapore	8.5	5.6	1.9	7.8	14.3	1.1
Simple average	6.4	4.1	1.2	6.7	13.5	1.1
Weighted average	7.4	4.8	1.4	7.7	15.3	—

— Not available.
Source: World Bank 1994a (table A.1).

Types of pension provision and pension administration

Country and type of state	Pension provision	Pension administration
Australia (Federal)	Commonwealth (central) government (Uniform).	Central and state governments.
Austria (Federal)	Federal government (Uniform).	Federal government.
Canada (Federal)	Three pillars: • Organization of American States (Universal flat-rate pension to everyone aged 65 and over.) • Canada and Quebec pension plans (partially funded.) • Guaranteed Income Supplement (GIS) and similar provincial supplements (OAS-GIS are pay as you go and six provincial governments have special "top up" for poorer pensioners.)	Federal and provincial governments (Canada and Quebec have legally separated pension plans but they are similar in contribution and benefits. There is full portability between the two.)
Chile (Unitary)	Fully funded system, guaranteeing the minimum pension (Originally fragmented in early 1970s, gradually unified and reformed in 1970–80.)	Decentralized (Privately managed AFPs are regulated by Pension Administration, an agency of the Ministry of Labor and Social Security.)
Colombia (Unitary)	Fragmented (There were 1040 institutions in 1990, and only 20.9% of population was covered.)	Decentralized
France (Unitary)	Central government (Unified, supplemented by mandatory occupational programs and minimum pension guarantees.)	National old-age insurance fund (CNAV) (Centrally managed, administered by a council of 25 members.)
Germany (Federal)	Separate programs for white-collar and blue-collar, civil servants, railway workers, seamen, miners, craftsmen, agricultural workers, and self-employed. (Unified policy formation but fragmented provision. It is a tax-financed pay as you go system.)	Fragmented (blue-collar insurance program is controlled by 23 state insurance institutions (LVA)–independent corporations governed by an assembly consisting of government officials and representatives of employers and workers.)
India (Unitary)	The Provident Fund Scheme, the Family Pension Scheme, and the Gratuity Scheme (Unified, but covers only the formal sector).	Ministry of Labor (Provident and Family Pension Funds). Boards of trustees at the central and state levels manage the funds. Gratuity Funds are administered by state governments and labor commissioners.
Indonesia (Federal)	Programs for government employees, armed forces, and private- and state-enterprises workers are under ASTEK. Income protection for old-age is provided by a provident fund.	Federal government (responsible for administering the nation's social insurance system.) PERUM ASTEK (a public enterprise that operates and administers the programs.)
Israel (Unitary)	Under the National Insurance Law of 1953 every resident aged 18 or over is insured and must contribute (Unified 3-pillar system).	National Insurance Institute (Centrally administered.)
Japan (Unitary)	Universal flat basic pension at age 65 under NPI and employees can continue to receive earnings-related pensions from old systems. Unified and two-pillar system introduced in 1985, fragmented system before 1985.	Social Insurance Agency (Centrally administered at the national level) Insurance division of prefecture welfare department and local SIA branches (local administration) The Ministry of Health and Welfare (operation of all public pension programs).
Korea, Rep. of (Unitary)	National comprehensive pension program covering the entire workforce. (The programs for public employees, military, and school teachers remain separated. Implemented in 1993. Initially fragmented with a small coverage.)	Multiple government ministries (centrally administered causing many administrative conflicts and much inefficiency.)

Types of pension provision and pension administration

Country and type of state	Pension provision	Pension administration
Malaysia (Federal)	Unified provision.	Decisionmaking on social security is centralized. The system is administered by a tripartite governing board, Employees' Provident Fund (EPF) (manages the old-age, survivors' and invalidity grants programs). The Ministry of Finance and Human Resources supervises the institution.
Mexico (Federal)	Mexican Social Security Institute (IMSS) (Unified, created under the Social Security Act of 1943. Fragmented and different social security schemes from one state to another before 1943 despite many attempts to unify them.)	Mexican Social Security Institute (IMSS) (Centrally administered)
Singapore (Unitary)	The Central Provident Fund (CPF) (compulsory savings scheme but offers no protection for low income earners. It is inequitable since the more one earns the more tax-free contribution is allowed, unified pension system.)	CPF (administration by a government investment agency–but investment is carried out by the Government of Singapore Investment Corporation (GSIC) and the Monetary Authority of Singapore (MAS).)
Sweden (Unitary)	Basic old-age pension (unified, given to all citizens regardless of contributions.) State earnings-related pension (ATP) (together it gives a replacement rate of about 65%.)	Social Insurance Board (RFV) (Centrally administered)
Switzerland (Federal)	National Old-age and Survivors Insurance program (AHV) (Unified covering the entire population); occupational retirement schemes and private savings.	Decentralized (A personal ID number is given to allow registration at one of the 107 "compensation institutions." The Central Income Substitution Office is in charge of settling accounts between the institutions and keeping a central register. The institutions and employers are inspected twice a year.)
United Kingdom (Unitary)	Initially fragmented, unified after the National Insurance Act of 1946. Two pillars include a flat-rate "basic component pension," and an earnings-related "additional component pension."	The Department of Health and Social Security (DHSS) (Responsible for the Social Security System in mainland Britain. There is a separate system in Northern Ireland. Administered by multiple government agencies until 1968.)
United States (Federal)	OASDI (Uniform), nine other programs are operated by the states. (Initially fragmented when 30 states enacted own laws. The Social Security Act of 1935 transferred the responsibility for income maintenance to the federal government. Federal uniformity of Old-Age Insurance (OAI) was achieved.)	Social Security Administration (SSA) in the Department of Health and Human Services (Central administration of the OASDI system; SSA has over 1300 district offices, prepares actuarial cost estimates and monitors financial soundness of the system. The Treasury Department collects payroll taxes through IRS, makes payments, and manages funds.)

Note: Because of space limitations, this table focuses on the main mandatory pension system for countries with multipillar systems.
Source: Compiled based on Wang 1995.

TABLE A2.2
Household income comparisons, United States and China

United States	Median household income, 1989 (U.S.$)	United States	Median household income, 1989 (U.S.$)	China, province	Urban household disposable income 1993 (Yuan)
Maine	27,854	South Carolina	26,256	Hebei	3,224
New Hampshire	36,329	Georgia	29,021	Shanxi	1,957
Vermont	29,792	Florida	27,483	Inner Mongolia	1,893
Massachusetts	36,952	Kentucky	22,534	Liaoning	2,314
Rhode Island	32,181	Tennessee	24,807	Jilin	1,953
Connecticut	41,721	Alabama	23,597	Heilongjiang	1,960
New York	32,965	Mississippi	20,136	Jiangsu	2,774
New Jersey	40,927	Arkansas	21,147	Zhejiang	3,626
Pennsylvania	29,069	Louisiana	21,949	Anhui	2,448
Ohio	28,706	Oklahoma	23,577	Fujian	2,839
Indiana	28,797	Texas	27,016	Jiangxi	1,919
Illinois	32,252	Montana	22,988	Shandong	2,515
Michigan	31,020	Idaho	25,257	Henan	1,963
Wisconsin	29,442	Wyoming	27,096	Hubei	2,450
Minnesota	30,909	Colorado	30,140	Hunan	2,688
Iowa	26,229	New Mexico	24,087	Guangdong	4,632
Missouri	26,362	Arizona	27,540	Guangxi	2,895
North Dakota	23,213	Utah	29,470	Hainan	3,072
South Dakota	22,503	Nevada	31,011	Sichuan	2,421
Nebraska	26,016	Washington	31,183	Guizhou	2,313
Kansas	27,291	Oregon	27,250	Yunnan	2,653
Delaware	34,875	California	35,798	Tibet	—
Maryland	39,386	Alaska	41,408	Shaanxi	2,102
District of Columbia	30,727	Hawaii	38,829	Gansu	2,003
Virginia	33,328			Qinghai	2,127
West Virginia	20,795			Ningxia	2,171
North Carolina	26,647			Xinjiang	2,423
Mean			29,134.67		2,512.88
Standard deviation			5,509.21		622.42
Coefficient of variation			0.19		0.25

— Not available.
Source: Statistical Abstract of the United States 1993 (p. xix); *China Statistical Yearbook 1994* (table 9-13, p. 265).

TABLE A2.3
Share of recent average earnings paid as a pension after thirty years of covered employment, selected countries, 1991
(percent)

Region and country	Replacement rate	Region and country	Replacement rate
Latin America and the Caribbean		*Europe*	
Argentina	70	Austria	57
Ecuador	75	Czech Republic	55
Guatemala	70	Portugal	66
Honduras	65	Spain	90
Mexico	60	Turkey	70
Panama	88	Yugoslavia, Federal Republic[a]	65
Paraguay	67		
Peru[a]	80	*South Asia*	
St. Lucia	60	Pakistan	60
Uruguay[a]	60		

a. Data for men only.
Source: World Bank 1994a.(table A.8, p. 369).

Contribution rates for social security programs—OECD countries, 1995
(percent)

Country	Old age, disability, death			All social security programs		
	Insured person	Employer	Total	Insured person	Employer	Total
Australia	0	0	[b]	1.25	0	1.25[b]
Austria	10.25	12.55	22.80	17.20	25.30	42.50
Belgium	7.50	8.86	16.36	13.07	27.44	40.51
Canada	2.70	2.70	5.40	5.70	8.40	14.10[c]
Denmark	[d]	[d]	[d]	[d]	[d]	[c, d]
Finland	5.55	19.00	24.55	7.45	22.00	29.45
France	8.05	8.20	16.25	18.27	34.31	52.58
Germany	9.30	9.30	18.60	16.55	17.99	34.54[c]
Greece	6.67	13.33	20.00	11.95	23.90	35.85
Iceland	4.00	8.50	12.50	4.00	15.75	19.75
Ireland	7.75[e, f]	12.20[e, f]	19.95[e, f]	7.75[f]	12.20[f]	19.95[c, f]
Italy	8.34	21.30	29.64	9.34	47.62	56.96
Japan	8.25	8.25	16.50	12.24	14.38	26.62
Luxembourg	8.00	8.00	16.00	15.00	13.00	28.00
Netherlands	25.78	0	25.78	38.93	10.75	49.68[c]
New Zealand	0	0	[b]	0.80	1.85	2.65
Norway	7.80[e]	14.20[e]	22.00[e]	7.80	14.20	22.00[c]
Portugal	11.00	23.75[h]	34.75[h]	11.00	26.75	37.75
Spain	4.70	23.60[i]	28.30[i]	6.30	32.00	38.30
Sweden	1.00	19.03	20.03	3.95[j]	30.96	34.91[c]
Switzerland	4.90	4.90	9.80	6.40	7.74	14.14
Turkey	9.00	11.00	20.00	14.00	19.50	33.50
United Kingdom	12.00[e, f]	10.20[e, f]	22.20[e, f]	12.00[f]	10.20[f]	22.20[c, f]
United States	6.20	6.20	12.40	7.65	13.35	21.00

a. Includes old age, disability, death: sickness and maternity; work injury; unemployment; and family allowances. In some cases, only certain groups, such as wage earners, are represented.
b. The central government pays the entire cost of most programs from general revenues.
c. The central government pays the whole cost of family allowances.
d. Portion of set amount for old age, disability, death, central and local government, and other types of contributions for the other programs.
e. Also includes rate for sickness and maternity, work injury, and unemployment.
f. Range according to earnings bracket. Higher rate is shown, which applies to highest earnings class.
g. Plus flat amount for work injury.
h. Also includes rate for sickness and maternity, unemployment and family allowances.
i. Also includes rate for sickness and maternity and family allowances.
j. Plus flat amount for unemployment.
Source: SSA 1996.

Real rate of return to private pension funds in industrial countries, 1970–90

(percent)

Country	1970–75	1975–80	1980–85	1985–90	Average 1970–90
Canada	−1.8	−1.1	5.2	7.7	2.2
Denmark	−2.0	0.8	16.9	—	4.1
Germany	3.3	3.2	7.6	6.2	5.1
Japan	−1.0	−1.6	10.9	13.6	4.4
Netherlands	−1.5	1.9	10.4	6.2	4.2
Switzerland	−1.4	3.7	2.7	−0.2	1.2
United Kingdom	−0.5	5.0	12.4	8.0	6.1
United States	−1.6	−2.0	7.7	9.6	3.3
Average	−0.8	1.2	9.2	7.3	3.8

— Not available.

Note: These rates of return are based on portfolio allocations among asset types, and annual data on yields and prices by asset types. Average is unweighted simple average.

Source: Davis 1995; World Bank 1994a (p. 176).

Chile: Real rate of returns and operating costs for pension funds

(percent)

Year	Real rates of returns	Operating costs
1981	12.5	—
1982	26.8	14.3
1983	22.7	7.3
1984	2.8	5.6
1985	13.4	4.1
1986	12.0	3.4
1987	6.4	2.9
1988	4.7	2.9
1989	6.6	2.8
1990	17.6	2.3
1991	28.6	1.8
1992	4.0	1.9
1993	16.7	—
1994	17.8	—
Average	14.0	

— Not available.

Source: Jeraldo and Munoz 1995 (p. 141).

Malaysia: Employees Provident Fund investment portfolio mix, 1987–94

(percentage shares)

Assets	1987	1988	1989	1990	1991	1992	1993	1994
Government securities (MGS)	89.3	88.1	84.7	79.2	73.6	65.1	55.1	48.7
Money market instruments	1.0	0.7	7.6	10.5	13.3	19.8	28.6	26.4
Debenture and loan[a]	6.2	5.2	5.7	8.2	11.0	12.4	12.1	14.6
Equities	1.9	1.8	1.5	1.6	1.6	2.2	3.4	10.2
Others[a]	1.6	4.1	0.5	0.6	0.5	0.5	0.7	0.1
Total investment	100.0	100.0	100.0	100.0	100.0	100.0	100.0	100.0
Amount (M$ million)	31,266	35,433	40,250	45,642	51,997	60,863	71,273	82,696
Real rate of return	7.7	5.4	5.2	5.4	3.6	3.3	4.4	4.2

a. Indicates promissory notes, debenture loan, and corporate bonds.
b. Indicates portfolio managers, advanced deposits, and other securities.

Source: Asher 1995 (p. 46).

Estimated rates of return to capital and ratios of pretax profits to total assets
(1985 output value)

Year	Industry	Construction	Transportation	Commerce	Ratio of pretax profit to total assets
1952	0.3065	1.1606	0.0890	0.2100	26.73
1953	0.2981	0.8827	0.0890	0.1633	32.00
1954	0.2872	0.7490	0.0850	0.1235	32.00
1955	0.2654	0.7282	0.0819	0.1057	31.57
1956	0.2469	0.8515	0.0779	0.1029	33.78
1957	0.2368	0.6628	0.0910	0.0933	36.41
1958	0.3107	1.0388	0.1092	0.1647	48.93
1959	0.2461	0.9749	0.0900	0.1565	51.25
1960	0.2318	0.6955	0.0850	0.1592	45.88
1961	0.2058	0.5543	0.0738	0.0700	16.73
1962	0.1873	0.4399	0.0668	0.0700	15.89
1963	0.1822	0.4503	0.0678	0.0672	21.57
1964	0.1797	0.4622	0.0668	0.0645	27.04
1965	0.1797	0.4711	0.0647	0.0618	31.36
1966	0.1772	0.4651	0.0627	0.0576	36.30
1967	0.1755	0.4547	0.0617	0.0563	22.83
1968	0.1738	0.4533	0.0637	0.0549	18.20
1969	0.1764	0.4443	0.0617	0.0508	26.62
1970	0.1797	0.4354	0.0587	0.0480	32.20
1971	0.1822	0.4221	0.0566	0.0467	31.57
1972	0.1814	0.4072	0.0536	0.0439	29.15
1973	0.1789	0.3908	0.0506	0.0412	27.15
1974	0.1772	0.3775	0.0496	0.0398	22.83
1975	0.1772	0.3686	0.0485	0.0412	23.89
1976	0.1772	0.3611	0.0475	0.0425	20.31
1977	0.1747	0.3641	0.0465	0.0412	23.21
1978	0.1831	0.3552	0.0445	0.0357	26.45
1979	0.1806	0.3492	0.0445	0.0357	26.70
1980	0.1806	0.3373	0.0445	0.0343	26.50
1981	0.1789	0.3284	0.0435	0.0357	25.31
1982	0.1772	0.3225	0.0425	0.0357	24.93
1983	0.1747	0.3210	0.0425	0.0357	24.71
1984	0.1738	0.3255	0.0435	0.0412	25.42
1985	0.1705	0.3195	0.0435	0.0480	23.80

Source: Yu 1995 (columns 1–4 are calculated from table XIV in Chow 1993; column 5 is calculated from table 10–22, p. 437, *China Statistical Yearbook 1993*).

TABLE A4.4b

Pretax profits to total assets
(1985 output value)

Year	Ratio of pretax profits to total assets	Year	Ratio of pretax profits to total assets
1986	20.06	1991	8.63
1987	18.77	1992	9.70[a]
1988	17.43	1993	10.30[a]
1989	13.08	1994	9.77[a]
1990	9.35	1995	8.29[a]

a. Indicates that the number is not in 1985 output value.
Source: Yu 1995 (table 10–22); *China Statistical Yearbook 1994*.

TABLE A4.5
Interest rates of major savings deposits of enterprises and institutions, 1990–94

Item	4/15/90	8/21/90	4/21/91	5/15/93	7/11/93	1994–95	5/1996
Demand deposits	2.88	2.16	1.8	2.16	3.15	3.15	2.97
Term deposits							
3 months	6.30	4.32	3.24	4.86	6.66	6.66	4.86
6 months	7.74	6.48	5.40	7.20	9.00	9.00	7.20
1 year	10.08	8.64	7.56	9.18	10.98	10.98	9.18
2 years	10.98	9.36	7.92	9.90	11.70	11.70	9.90
3 years	11.88	10.08	8.28	10.8	12.24	12.24	10.80
5 years	13.68	11.52	9.00	12.06	13.86	13.86	12.06
8 years and over	16.20	13.68	10.08	14.58	17.10	17.10	

Source: China Statistical Yearbook 1995 (pp. 576–577); World Bank 1995e.

TABLE A4.6
Interest rates of major loans, 1990–94

Item	April 1990	August 1990	April 1991	May 1993	July 1993	1994–95	5/1996
Loans for working capital							
3 months	7.92	7.92	—	—	—	—	
6 months	9.00	8.64	8.10	8.82	9.00	9.00	9.72
1 year	10.08	9.36	8.64	9.36	10.98	10.98	10.98
Loans for private businesses							
Base rate	10.08	9.36	8.64	9.36	10.98	10.98	
Floating rate (x%) over the base	+20	+20	+20	+20	+20	+20	
Loans for technical innovation	10.08	9.36	8.46	9.18	10.98	10.98	
Loans for capital construction							
1 year and less	10.08	9.36	8.46	9.18	10.98	10.98	11.52
1–3 years	10.80	10.08	9.00	10.80	12.24	12.24	13.14
3–5 years	11.52	10.80	9.54	12.06	13.86	13.86	14.94
5–10 years (5+ years after 1991)	11.88	11.16	9.72	12.24	14.04	14.04	15.12
Over 10 years (before 1991)	11.88	11.16	a				

— Not available.
a. In 1991 the two categories merged into one, 5+ years.
Source: China Statistical Yearbook 1995 (pp. 576–577); World Bank 1995e.

TABLE A4.7
Bond interest rates

Treasury bonds			*For individuals (indexed)*	Ignored
6 months (1994)	9.80		For units	
1 year (1994)	10.96		3 years (1993)	13.96
2 years (1994)	13.00		5 years (1993)	15.86
3 years (1994)	13.98 (indexed)			
Fiscal bonds			5 years (1991)	9.00
2 years (1988)	8.00		(1992)	9.50
5 years (1990)	10.00		(1993)	14.00
Financial bonds				
1 year (1992)	8.50		4 years (1992)	10.50
2 years(1992)	9.00		5 years (1989)	10.50
3 years (1992)	9.50			

Source: World Bank 1995e.

TABLE A4.8
Asset allocation of pension funds in industrial countries, 1990
(percentage of assets)

Instruments or assets	United Kingdom	United States	Germany	Japan	Canada	Switzerland
Short-term assets	7	9	2	3	11	12
Market papers	1	3	—	—	10	10
Deposits	6	6	2	—	1	1
Bonds	14	36	25	47	47	29
Government bonds	11	20	17	5	39	—
Private bonds	3	16	8	—	8	—
Equities	63	46	18	27	29	16
Mortgages	—	2	9	1	4	8
Loans	0	0	36	13	0	14
Property	9	—	6	2	3	17
Foreign assets	18	4	1	7	6	5
Other assets	6	2	1	—	2	1
Total	117	99	98	100	102	102

— Not available.
Note: There are some overlapping categories (equity and foreign equity) in some countries, so they do not add up to unity.
Source: Davis 1995 (p. 161).

TABLE A4.9
Real returns and risks by asset types, 1967–90

	United Kingdom		United States		Germany		Japan		Canada		Switzerland	
Loans	1.4	(5.0)	3.5	(2.9)	5.3	(1.9)	0.9	(4.3)	4.0	(3.7)	2.6	(2.0)
Mortgages	2.0	(5.2)	2.0	(13.4)	4.7	(1.4)	3.0	(4.9)	2.4	(12.3)	1.3	(2.3)
Equities	8.1	(20.3)	4.7	(14.4)	9.5	(20.3)	10.9	(19.4)	4.5	(16.5)	6.2	(22.3)
Bonds	−0.5	(13.0)	−0.5	(14.3)	2.7	(14.9)	0.2	(12.8)	0.0	(12.1)	−2.2	(17.6)
Short-term assets	1.7	(4.9)	2.0	(2.5)	3.1	(2.1)	−0.5	(4.6)	2.5	(3.3)	1.2	(2.2)
Property	6.7	(11.4)	3.4	(6.4)	4.5	(2.9)	7.2	(6.8)	4.6	(6.2)	3.7	(8.9)
Foreign bonds	−0.1	(15.0)	1.6	(14.9)	3.0	(11.2)	1.3	(14.6)	−1.7	(12.7)	−1.7	(12.6)
Foreign equities	7.0	(16.2)	9.9	(17.2)	10.4	(13.5)	7.8	(18.7)	5.8	(14.3)	5.6	(16.0)

Note: Numbers in parentheses are standard deviations.
Source: Davis 1995 (p. 133).

TABLE A4.10

Real rate of return and total assets for pension funds, 1983–93

Country	Real rate of return (percent per year)	Total assets end of 1993 ($ billion)
France	13.0	199.7
Australia	12.0	122.2
Ireland	11.8	13.8
United Kingdom	11.5	726.4
New Zealand	10.8	7.9
United States	9.5	2,908.0
Belgium	9.0	5.6
Spain	8.8	13.4
Canada	8.3	162.3
Denmark	8.2	14.4
Hong Kong	8.1	6.8
Norway	8.0	30.6
Netherlands	7.5	216.2
Sweden	7.2	61.8
Germany	6.5	254.2
Japan	6.2	1,752.7
Switzerland	3.5	195.3
Singapore	1.8	32.9

Source: "UBS Asset Management," *The Economist,* 8 October 1994, p. 123.

TABLE A4.11

Insurance premiums as a share of GNP, 1985–94
(billions of current yuan)

	1985	1991	1992	1993	1994	Annual growth
Total premiums	2.57	15.57	33.51	45.68	37.64	136.5
Pension insurance	0.18	2.33	5.31	3.53	3.32	174.4
Other (including life insurance)	—	1.30	8.03	14.52	8.52	138.8
Nominal GNP	899.40	2,166.50	2,665.10	3,447.60	4,491.80	
Ratio of premiums to GNP (percent)	0.29	0.72	1.26	1.32	0.84	19.0

— Not available.
Source: World Bank staff estimates calculated from *China Statisitcal Yearbook 1995* (p. 580).

Asset allocation in the People's Insurance Company of China
(millions of yuan and percent)

	1989	1990	1991	1992	1993
Liquid assets (cash and deposits)	1,059.6	1,500.4	1,924.6	2,220.5	2,307.4
Investment	381.7	509.4	757.5	1,013.6	1,558.7
Total assets[a]	1,741.4	2,384.2	3,515.8	4,800.8	5,851.8
Ratios					
Liquid assets to total assets	60.8	62.93	54.74	46.25	39.43
Investments to total assets	21.9	21.37	21.55	21.11	26.64
Investments to liquid assets	36.0	33.95	39.36	45.65	67.56

a. Total assets include some items not listed.
Source: Yu 1995; *Almanac of China's Finance and Banking 1994.*

FIGURE A4.1

Asset distribution of U.S. life insurers, 1991

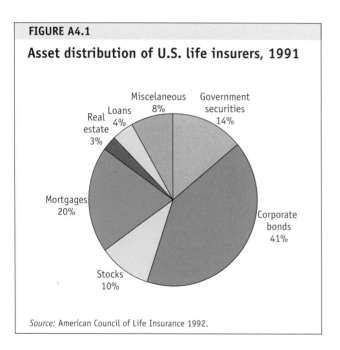

Source: American Council of Life Insurance 1992.

References

Ahmad, Ehtisham, and Athar Hussain. 1991. "Social Security in China: A Historical Perspective." In E. Ahmad and A. Hussain, eds., *Social Security in Developing Countries*. Oxford: Oxford University Press.

American Council of Life Insurance. 1992. *Life Insurance Fact Book Update*. Washington, D.C.

Asher, Mukul. 1995. "Investing National Provident Fund Balances in Malaysia and Singapore: Practices and Issues." A background paper prepared for the World Bank, China and Mongolia Department, Washington, D.C.

Auerbach, A.J., and L.J. Kotlikoff. 1985. "Simulating Alternative Social Security Responses to Demographic Transition." *National Tax Journal* 38: 153–68.

Barr, Nicholas. 1994. *Labor Markets and Social Policy in Central and Eastern Europe: The Transition and Beyond*. New York: Oxford University Press.

Blinder, Alan S. 1988. "Why is the Government in the Pension Business?" In Susan M. Wachter, ed., *Social Security and Private Pensions: Providing for Retirement in the Twenty-First Century*. Lexington, Mass.: Lexington Books.

Borchard, Dagmar. 1993. "Legal Framework of Welfare Provisions and Social Security." In Renate Krieg and Monika Schadler, eds., *Social Security in the People's Republic of China*. Hamburg: Mitteilungen Des Instituts Fur Asienkunde.

Botka, Andras. 1994. "Some Features of Current Pension System Reform in Latin America." *Revista de Analisis Economico* 9(1): 211–35

China Labor Yearbook Editorial Board. 1995. *China Yearbook of Labor Statistics 1994*. Beijing: China Labor Press.

China State Statistical Bureau. Various years. *China Statistical Yearbook*. Beijing: China Statistical Press.

CHIRD (China Institute for Reform and Development). 1993. "Guiding Thoughts of China's Social Security System Reform and Reform Programs of Hainan Province." Paper presented at the International Symposium on the Theoretical and Practical Issues of the Transition toward the Market Economy in China, July, Haikou.

Chow, Gregory C. 1993. "Capital Formation and Economic Growth in China." *Quarterly Journal of Economics* 108(3): 809–42.

Chow, Nelson W.S. 1988. *The Administration and Financing of Social Security in China.* Hong Kong: University of Hong Kong, Centre of Asian Studies.

Davis, E. Philip. 1993. "The Structure, Regulation, and Performance of Pension Funds in Nine Industrial Countries." Policy Research Working Paper 1229. World Bank, Financial Sector Development Department, Washington, D.C.

———. 1995. *Pension Funds, Retirement-Income Security, and Capital Markets: An International Perspective.* Oxford: Oxford University Press.

Friedman, Barry, and Leonard Hausman. 1994. "Social Protection Policy and Economic Restructuring in China." Brandeis University, Department of Economics, Waltham, Mass.

Gong, Seng, and Zicheng Yie. 1995. "A Research Report on the Basic Pension Insurance for Enterprise Employees, and Comparison Between Plan I and II." Ministry of Labor, Institute of Social Insurance, Beijing.

Gorsetti, G., and K. Schmidt-Hebbel. 1995. "Pension Reform and Growth." Policy Research Working Paper 1471. World Bank Policy Research Department, Washington, D.C.

Groves, Theodore, Yongmiao Hong, John McMillan, and Barry Naughton. 1994. "Autonomy and Incentives in Chinese State Enterprises." *Quarterly Journal of Economics* 10 (February): 183–209.

———. 1995. "China's Evolving Managerial Labor Market." *Journal of Political Economy* 103(4): 873–92.

Guo, Shuqing. 1995. "The Restructuring of State Assets and the Compensation of the Social Security Debts." Background paper prepared for the World Bank, China and Mongolia Department, Washington, D.C.

Holtzman, Robert. 1994. "Funded and Private Pensions for Eastern European Countries in Transition?" *Revista de Analisis Economico* 9(1):183–210.

Hu, Zu-liu. 1994. "Social Protection, Labor Market Rigidity, and Enterprise Restructuring in China." IMF Paper on Policy Analysis and Assessment. International Monetary Fund, Fiscal Affairs Department, Washington, D.C.

Hussain, Athar. 1993. "Reform of the Chinese Social Security System." In E. Ahmad and A. Hussain, eds., *Social Security in the People's Republic of China.* Oxford: Oxford University Press.

ILO (International Labour Organisation). 1989. *Social Security Protection in Old Age.* Geneva: International Labour Office.

———. 1994. *Yearbook of Labor Statistics.* Geneva: International Labor Office.

———.1995. *World Labor Report.* Geneva: International Labour Office.

IMF (International Monetary Fund). 1996a. "Aging Population and the Fiscal Consequences of Public Pension Schemes with Particular Reference to the Major Industrial Countries." Fiscal Affairs Department, Washington, D.C.

———. 1996b. "Pension Regimes and Saving," Fiscal Affairs Department, Washington, D.C.

James, Estelle. 1997a. "Pension Reform: Is there a Tradeoff between Efficiency and Equity?" Policy Research Working Paper 1767. World Bank, Policy Research Department, Washington, D.C.

———. 1997b. "New Systems for Old Age Security." Policy Research Working Paper 1766. World Bank, Policy Research Department, Washington D.C.

Jeraldo, Julio B., and Osvaldo M. Munoz, eds. 1995. *The Chilean Pension System Based on Individual Capitalization.* Santiago, Chile: Superintendency of Pension Funds Administrators.

Kotlikoff, Laurence J., and Jan Walliser. 1995. "Applying Generational Accounting to Developing Countries." Working Paper. Boston University.

Krieg, Renate. 1993. "The 'Preliminary Regulations for Social Security of Hainan Province'." In E. Ahmad and A. Hussain, eds., *Social Security in the People's Republic of China.* Oxford: Oxford University Press.

Lee, Chingboon. 1994. "China in Transition: Reform of the Urban Social Security System." World Bank Resident Mission, Beijing.

Mesa-Lago, Carmelo. 1994. *Changing Social Security in Latin America: Toward Alleviating the Social Costs of Economic Reform.* Boulder: Lynne Rienner Publishers.

PICC (The People's Insurance Company of China). 1994. *The People's Insurance Company of China Annual Report 1993.* Beijing.

RGSSS (Research Group for the Social Security System), Government of China. 1995. "Projects for Primary Pension Insurance for Urban Staff and Workers." Beijing.

Schmahl, Winfried. 1993. "Harmonization of Pension Schemes in Europe? A Controversial Issue in the Light of Economics." In A.B. Atkinson and Martin Rein, eds., *Age, Work and Social Security.* New York: St. Martin's Press.

Schmidt-Hebbel, K. 1995. *Colombia's Pension Reform: Fiscal and Macroeconomic Effects.* World Bank Discussion Paper 314. Washington, D.C.: World Bank

Schmidt-Hebbel, K., and L. Serven. 1996. *Saving across the World: Puzzles and Policies.* World Bank Discussion Paper 359. Washington, D.C.: World Bank

Shao Lei, and Chen Xiangdong. 1991. *Zhongguo shehui baozhang zhidu gaige* [China's Social Security Reform]. Beijing: Economic Management Press.

Shilling, John D., and Yan Wang. 1996. "Managing Capital Flows in East Asia." World Bank, East Asia and Pacific Region, Washington, D.C.

SRC (State Reform Commission). 1995. *Shehui baozhang tizhi gaige* [Reform of Social Security System]. Beijing: Reform Press.

Sreuerle, C. Eugene, and Jon M. Bakija. 1994. *Retooling Social Security for the 21st Century: Right and Wrong Approaches to Reform.* Washington, D.C.: The Urban Institute Press.

SSA (Social Security Administration). 1995. *Social Security Programs throughout the World—1995.* SSA Publication 13-11805. Washington, D.C.: U.S. Government Printing Office.

Verbon, Harrie. 1988. *The Evolution of Public Pension Schemes.* Berlin: Springer-Verlag.

Vittas, Dimitri. 1995a. "Pension Funds and Capital Markets." Working Paper. World Bank, Policy Research Department, Washington, D.C.

———. 1995b. "Sequencing Social Security, Pension, and Insurance Reform." Policy Research Working Paper 1551. World Bank, Fiscal Sector Development Department, Washington, D.C.

———. 1995c. "Strengths and Weaknesses of the Chilean Pension Reform." World Bank, Policy Research Department, Washington D.C.

———. 1996. "Private Pension Funds in Hungary: Early Perfomance and Regulatory Issues." Policy Research Working Paper 1638. World Bank, Fiscal Sector Development Department, Washington, D.C.

Vittas, Dimitri, and Augusto Iglesias. 1992. "The Rationale and Performance of Personal Pension Plans in Chile." Policy Research Working Paper 867. World Bank, Financial Sector Development Department, Washington, D.C.

Wang, Shaoguang. 1995. "China: Pension Provision and Pension Administration." Background paper prepared for the World Bank, China and Mongolia Department, Washington, D.C.

Wong, Christine P. W., Christopher Heady, and Wing T. Woo. 1995. *Fiscal Management and Economic Reform in the People's Republic of China.* Oxford: Oxford University Press.

World Bank. 1990. *China: Reforming Social Security in a Socialist Economy.* Washington, D.C.

———. 1994a. *Averting the Old Age Crisis: Politics to Protect the Old and Promote Growth.* Wold Bank Policy Research Series. New York: Oxford University Press.

———. 1994b. *World Population Projections 1994–1995.* Washington, D.C.

———. 1995a. *China: The Emerging Capital Market.* Washington, D.C.

———. 1995b. *China: Reform of State-Owned Enterprises.* Washington, D.C.

———. 1995c. *The Emerging Asia Bond Market.* Washington, D.C.

———. 1995d. *Infrastructure Development in East Asia and Pacific: Toward a New Public-Private Partnership.* Washington, D.C.

———. 1995e. *Interest Rate Policy in China: From Rationalization to Liberalization.* Washington, D.C.

———. 1995f. *World Debt Tables 1995.* Washington, D.C.

———. 1996. "China: Pension System Reform." Report 15121-CHA. East Asia and Pacific Region, China and Mongolia Department, August, Washington, D.C.

Yao, Hong, and Jindou Wang. 1995. "China's Life Insurance Market and the Establishment of Enterprise Supplementary Pension Insurance System." Background paper prepared for World Bank, China and Mongolia Department, Washington, D.C.

Yu, Mingde. 1995. "Background Paper on Pension Fund Administration and Investment Policy in China." Background paper prepared for World Bank, China and Mongolia Department, Washington, D.C.

Zhou Chuanye. 1993. "Reform of Pension Insurance in Hainan Province." In Renate Krieg and Monika Schadler, eds., *Social Security in the People's Republic of China.* Hamburg: Mitteilungen Des Instituts Fur Asienkunde.

Zuo, Xuejin. 1995. "Transition of the Pension System from a Unfunded System to a Funded System and Society Debt: The Case of Shanghai." Background paper prepared for World Bank, China and Mongolia Department, Washington, D.C.